Practical Benchmarking: The Complete Guide

AF813037

Practical Benchmarking: The Complete Guide

Mohamed Zairi

Unilever Lecturer in TQM
Bradford University

and

Paul Leonard

Senior Consultant
Xerox Quality Solutions
Rank Xerox Ltd

SPRINGER-SCIENCE+BUSINESS MEDIA, B.V.

© 1996 Springer Science+Business Media Dordrecht
Originally published by Chapman & Hall in 1996
Softcover reprint of the hardcover 1st edition 1996

Commissioned and produced by Technical Communications (Publishing) Ltd.

ISBN 978-94-010-4560-5 ISBN 978-94-011-1284-0 (eBook)
DOI 10.1007/978-94-011-1284-0

Apart from any fair dealing for the purposes of research or private study, or criticism or review, as permitted under the UK Copyright Designs and Patents Act, 1988, this publication may not be reproduced, stored, or transmitted, in any form or by any means, without the prior permission in writing of the publishers, or in the case of reprographic reproduction only in accordance with the terms of the licences issued by the Copyright Licensing Agency in the UK, or in accordance with the terms of licences issued by the appropriate Reproduction Rights Organization outside the UK. Enquiries concerning reproduction outside the terms stated here should be sent to the publishers at the London address printed on this page.

The publisher makes no representation, express or implied, with regard to the accuracy of the information contained in this book and cannot accept any legal responsibility or liability for any errors or omissions that may be made.

A Catalogue record for this book is available from the British Library

Contents

Foreword

by Bob Camp

The business improvement topic and quality tool called benchmarking is becoming widely understood and broadly applied. There are now applications in almost all segments of the economy including industrial firms that either produce a product or a service, non-profit organizations such as healthcare, government and education. The approach is starting to spread around the globe with initiatives in Europe, Asia Pacific and South America. This is commendable and reassuring and must show that there is significant interest in the approach and that it works.

What is missing, however, are books and reference material that are not solely prepared in the US where benchmarking started. Theses would include examples of applications relevant to the local area and industries. They would include references to articles written about benchmarking appearing in local publications. In this fashion those interested would have near hand case histories of the use of benchmarking and therefore become encouraged to use the technique.

Zairi and Leonard have done the benchmarking community a real service by documenting the European view and application of benchmarking to a wide range of examples. But they have not stopped there. Their text includes treatment of a number of related facets of benchmarking that makes this a fairly thorough text.

The text does cover the basics.These include the origins of benchmarking, the strategy for its use, the types of benchmarking that are the basic comparisons that can be made and the appropriate approach to choosing partners and data collection. However, they also cover some of the more difficult obstacles to understanding; namely how benchmarking relates to performance measurements and how it is integrated into the other continuous improvement initiatives that organizations invariably find they are pursuing.

The text also covers a wide range of examples. These include applications in several different sectors of the economy: utilities, financial services, education, healthcare and several organization functional areas such as: human resources, marketing and R&D. In addition, the experiences of many European firms are covered, not the least of which are those of: Siemens, Rank Xerox, Diahatsu and Philips. The text gives the authors' view of how to avoid pitfalls and the future of benchmarking; this includes the focus on competitiveness and the need for global market driven superiority.

A unique feature of the work is a 'Who is who in Benchmarking' section. This does not exist in one comprehensively organized publication. The reader has a ready reference to organizations who have prominence in benchmarking, offer assistance in the pursuit of this improvement approach and who promote benchmarking. This offers invaluable reference for those organizations that are starting their benchmarking journey.

New and different perspectives are needed to expand the art and practice of benchmarking and I welcome them and commend this book to your reading.

Robert C. Camp
Rochester, NY, USA

Preface

The authors met for the first time some two and a half years ago at a seminar on benchmarking organized by the UK Chapter of the European Foundation for Quality Management. Our presentations were approached from different standpoints, yet we were both pleasantly surprised to see how similar were the core messages. One presentation was from the standpoint of the academic, the other from the standpoint of an experienced practitioner.

Over the ensuing months, we began to share more and more platforms at benchmarking conferences both in the UK and abroad and would spend many hours before, during and after conferences talking and developing our overall understanding of benchmarking from many different viewpoints.

Early in 1993 someone, we're not quite sure who, said that the combination of academic and practitioner was a very powerful one and that we should write a book on the subject of benchmarking.

At first we rejected the idea because there are so many books, articles, conferences, seminars and workshops on the subject that we felt there wasn't any room for more. However, one day we sat down and looked at the proposition more carefully. Was there really no room for any more literature? Was there really nothing we could contribute? Had it really all been said?

After a couple of hours of debate we decided to give it a try! We (and our wives!) were later to regret this decision as writing the book began to consume more and more time and effort.

However, the die was now cast, so the process of writing began. Imagine the scene – one author based in Bradford and travelling extensively, the other in the Midlands and also travelling extensively. One using one word processing technology, the other using a different, incompatible system. The sheer size of the logistics problem began to dawn and we began to wonder if we had bitten off more than we could chew and whether the book would ever be completed. In order to minimize the problem we allocated a 'lead' writer to each chapter with the intent of regular progress reviews and aimed at adding value to the lead writer's work.

The definition of 'regular' took on a whole new meaning. Half an hour here, an hour there, in Bradford, in London hotels, in Paris, over the telephone, by exchanges of faxes etc. Time began to hurtle by and progress seemed to be moving at a snails's pace. In spite of all the pressures of other priorities the book gradually took shape. As each chapter took the form of a first draft it was forwarded to the publisher and eventually all the initial work was completed. The rest, as they say, is history.

Our intent was to produce a complete guide to benchmarking from a practical standpoint which would help any company, whether large or small, expert practitioner or beginner to improve their results by a much better use of this most powerful of quality tools.

We believe we have succeeded in this task but ultimately it is up to you, the reader, to decide.

We have to thank Bob Camp of Xerox Corporation, whose book *Benchmarking – the search for industry best practices that lead to superior performance*, is a world best seller, for writing such an encouraging foreword.

Above all though, we have to thank our respective wives Alweena Zairi and Edna Leonard, without whose unstinting support, as more and more midnight oil was burnt, this book would never have seen the light of day.

Part One

1

Competing in a modern business environment: what does it mean?

1.1 Introduction

A crucial question in business today is whether the notion of competition in a modern business environment needs to be radically re-examined. Many variables play a role in determining competitive success; new ones are emerging all the time.

Expressions such as 'market-driven strategies', 'customer-based competitiveness' and 'time to market' highlight a sense of urgency and a new business outlook where winning strategies focus entirely on the market and the end customer, and less on internal operations, technologies, products and services.

Are we learning from the Japanese and their global supremacy? The Japanese have always focused on the customer. Their disciplined approach sets long-term corporate objectives and is less distracted by shareholder issues. The difference between the Western and Japanese approach has been explained by Turpin (1993) as follows:

> The most obvious source of potential conflict lies in the different ways of keeping score. Corporate objectives can vary considerably between Japanese and Western companies. Whilst the Japanese partner is managing the business so as to increase market share and the ratio of new products, and the Western partner is looking for a high ROI [return on investment], and maximizing the shareholders' return, troubles lie ahead. Japanese executives typically manage their businesses for long-term benefits with little attention paid to the shareholder whilst they consider, as one Japanese put it, that US firms have become the slaves of Wall Street.

Being competitive in the 1990s requires unprecedented strengths. Competitiveness is often the ability to determine rational capability (through strengths and weaknesses) and to fulfil customer needs.

Winning comes through innovation, uniqueness (differentiation), teaching rather than following and a culture of continuous improvement and learning. Benchmarking can act as a catalyst to success and superiority.

1.2 Market-driven strategies – a first requirement

Developments such as downsizing, reorganizing, repositioning, forming strategic alliances with other firms, focusing on core activities (nicheing) and revamping the business portfolio are common in many large organizations. Indeed, alert companies are coming to grips with the need to be much more dynamic and responsive in the marketplace.

Having a market-driven approach, therefore, is not merely adding incremental changes to existing cultures of competitiveness and old ways of managing operations; being a market-driven organization currently means:

1. reorganizing the business to focus entirely on the customer/market of operation;
2. deploying resources, strengths and areas of expertise only in core activities;
3. defining strategies, technological choice, and building the required capability of delivery through an accurate and thorough understanding of who the customer is and what the requirements are;
4. building competitive advantage through the ability to deliver uniqueness and excellence in products and services with true compassion – being first and good matters more than being last and best;
5. establishing superior standards of performance, and achieving customer satisfaction by continuous review, and subsequent action on those aspects of value-added activity which are most important from the customer's perspective; and
6. driving business competitiveness through innovation. Understanding the value of customer-driven innovation is vital for business survivability, so activities should focus on:
   - replacing the old with the new,
   - improving the existing products/services,
   - developing new products/services for unfulfilled needs, and
   - pioneering new technologies to build supremacy.

Developing market-driven strategies has to reflect a commitment to changing existing cultures of business, streamlining and focusing all activities for the purpose of adding value that would ultimately benefit the end customer. As Siemens' quality motto reflects: 'Quality is when your customers come back and your products do not.'

A market-driven organization must reflect radical changes in its culture. It is suggested that the required changes would include three stages (Naumann and Shannon, 1992).

1. **Bliss** Establishing an internal system dealing with quality deficiencies and gaps in quality performance. This stage reflects a reactive culture focusing on negative quality aspects such as waste and customer complaints.
2. **Awareness** Becoming aware that customer satisfaction is critically important. In addition to dealing with deficiencies, organizations reflecting this culture use a more pro-active approach to measure what is important to the customer through, for instance, customer satisfaction surveys.
3. **Commitment** Operating in a totally pro-active way through a continuous effort to capture the voice of the customer and translate it internally.

Moving from 'bliss to commitment' requires much effort and can only be achieved through performance rather than wishful thinking. Getting closer to the customer is far more than a cliché, a statement of hope, as Shapiro (1988) argues:

> I am convinced that the term 'market-oriented' represents a set of processes touching on all aspects of the company. It is a great deal more than the cliché 'getting close to the customer'.

1.3 Time-based competition

Expressions such as 'time to market', 'concurrent or simultaneous engineering', 'parallel working', 'integrated engineering' and 'forward engineering', are used to describe a new approach to competitiveness based on speed, quality, and being first in the marketplace. These expressions describe changes in existing cultures of work where working hard is no longer necessarily desirable.

Time-based competition expresses the need to:

- deliver products and services to the end customer faster than competitors;
- surprising competitors by continuously launching new products/services faster and in an innovative way which is difficult to imitate; and
- working in harmony with suppliers so that deliveries take place on time.

Effective time-based competitiveness means that all cycle times need to be tracked down, for example:

- right-first-time design – the translation of the physical/emotional need of the customer into tangible prototypes;
- make-to-market cycle time – the planning, scheduling, production and delivery of products and services; and

- innovation activity – the time it takes to receive feedback on customer satisfaction, assess future requirements and translate these into new innovations.

Essentially, time-based competition is a new culture where time-based performance measures are developed and continuously tracked in all activities and processes to identify and eliminate bottlenecks and to inject new learning that would result in faster methods of operation.

Time-based competition is not just an expression of hope. It is not about superimposing time-reduction targets on existing processes and methods of work. It has to result from high level commitment and a dedication to challenge existing methods by close scrutiny of all processes.

Benefits from time-based competition are numerous. Besides removing waste and optimizing value-added activity through crystallization, visibility and effectiveness of the whole delivery system, the following benefits can be achieved:

- an ability to innovate more quickly than competitors and offer new products and services with higher standards of quality and lower prices;
- an ability to discontinue existing lines which are poor performers through continuous replacement;
- an ability to leverage time savings for new developments;
- an ability to harness synergy through teamwork and crossfunctional involvement; and
- time cycles can be made faster and faster due to the sequential recycling and injection of new learning.

Time-based competition means that organizations can leapfrog and outfox their competitors regularly, as reported in the following example (Spanner *et al.*, 1993).

> The competition can get overwhelmed by this strategy, as happened to Yamaha where Honda introduced 113 new motorcycle models in just 18 months in the early 1980s.

Synchronicity (Starr, 1992) or incremental innovation is a practice that most Japanese companies excel at. It means continuously presenting the customer with a stream of new products and services yet using time savings effectively by synchronizing all activities to improve existing methods and inject new learning.

This means that quality standards are raised and a continuous improvement culture is in evidence. True leaders never rest. They believe that the more they learn, the more they realize how naive they were. For example, in 1985, it took Motorola three years to develop the cellular phone. By 1991 the time was reduced to less than 50% and the commitment is to reduce the development time to less than six months (Kotler and Stonich, 1991). Similarly, 3M, considered by many as the best example of a leading innovator, has managed to develop a culture which is time conscious. There is continuous encouragement and support for new ideas; new methods of work emphasize speed and the ability to make things happen quickly.

1.4 Being a global competitor – what does it mean?

The challenge of operating in a global environment does not just depend on transferring technologies, managing global supply sources and developing global distribution networks. Creating a global company means operating in a world without walls and the company itself without a country.

It means establishing a global corporate culture. This goes well beyond changing management structures and strategies. It requires a means of understanding the needs of customers globally and a knowledge of how competitors are addressing similar issues.

Being a global competitor means having the ability to:

- establish global efficiency through the transfer of technology, know-how and the effective deployment of resources – the sharing of best practice becomes important;
- be pro-active and to harness innovation activity globally to establish superiority, i.e. offering uniqueness and differentiated products and services worldwide; and
- ensure flexibility and responsiveness to deliver customer needs at a global level.

The abilities described above mean that global organizations can retreat, advance, integrate, split global operations according to the needs of their customers and in response to market changes and competitor threats. For example, the following companies have developed strategies that enable them to operate effectively in a global market.

- Matsushita Electric Company, a competitor of Philips, has developed a global strategy to counter the threats from Philips and to exploit the weaknesses in the latter's approach (Rhinesmith, 1991):
  - effective utilization of input from all subsidiaries for global management strategies;
  - closeness to markets ensuring all development work answers market needs; and
  - investigation of all key activities (i.e. development, marketing and manufacturing) to ensure smooth transfer of responsibility.
- ABB allocates worldwide responsibility for decision making in relation to key products to centres with the greatest degree of competence (Theuerkauf, 1991).

- Unilever developed a new structure moving away from local production to concentrated manufacturing. In 1973 it had 13 factories in 13 countries; by 1989, this was concentrated to four factories in four countries.
- Procter & Gamble has reorganized to focus more on product groups being managed across geographical boundaries.
- IBM shifted its global headquarters for communication systems from New York to London. It has also moved all its R&D activities for laptop and notebook PCs to Japan (Theuerkauf, 1991).

Becoming a global competitor means challenging existing cultures radically by tackling existing management systems and structures so that all processes are streamlined and a global value-added system is established. The exploitation of information technology (IT) becomes a powerful means for exchanging information and communicating effectively in a continuous manner. Tools such as notebook PCs, fax machines, intercontinental video conference facilities and electronic mail are very powerful means of communication.

An example of the power of global communications is Hewlett-Packard's (H-P) success in launching new products (Rhinesmith, 1991). It is reported that 50% of H-P's sales volume comes from products launched in the last 2–3 years. This is achieved through global R&D co-ordination and integration allowing H-P to access the best ideas worldwide bringing them to the most appropriate markets.

Global competitors must develop management executives capable of understanding global markets and managing various conflicts. They must develop global skills through global career planning and encouraging mobility and secondments, and developing global centres of excellence, through the exploitation of material resources, human resource strength, cultural strengths, etc.

As Rhinesmith (1991) argues:

> Developing a global corporate culture is for most corporations the last step in the globalization agenda. It is not just a matter of doing business internationally or even having subsidiaries abroad. Developing a global corporate culture involves forming the integrated values, mechanisms, and processes that allow a company to manage change successfully in a competitive global marketplace.

1.5 Competing through learning – the role of core competences

Core competences are all those skills and levels of knowledge and expertise which give organizations advantages in the competitive arena and enable them to sustain high standards of competitiveness through smooth or turbulent times. Core competences are the answer to the question, 'What do we excel at?' As such, they are supposed to give organizations unique capability and standards difficult to imitate by competitors.

Core competences often represent a blend of hard and soft skills capable of (Sterne, 1992):

- providing access to a wide variety of markets;
- adding value to the product or service that would benefit the end customer.

All successful global companies tend to have specific core competences, for example:

- Sony has a skill in miniaturization;
- American Airlines has expertise in IT;
- Corning has expertise in glass and ceramics.

Core competences have to be exploited to the full by developing stretch objectives: otherwise they will only achieve incremental gains. It is therefore not surprising to see the degree of emphasis organizations are placing on innovation and the need to harness people's talents. As Sterne (1992) puts it:

> Core competences are the muscles of the organization. They make it possible to accomplish much of the work that is carried out every day. But as any marathon runner knows, muscles are not enough to win a race. A high level of aspirations is also needed for remarkable achievement.

1.6 Competing through benchmarking – a proposed model

Future competitiveness depends on organizations getting close to their markets and developing responsiveness to market needs, and an ability to counter threats and to exploit opportunities.

The following model illustrates a dynamic approach. The model shows two stages representing **push forces** which are the ones organizations can control least, and **pull forces** which are the capability of response, i.e. those which can be effectively controlled.

Stage 1 – the demand

This stage represents all the necessary steps for scanning the business environment continuously to establish the level of demand for products and services. Benchmarking can be a very powerful tool in building a good understanding of market conditions. Forces which apply pressure include (Fig. 1.1):

- customer
- global markets
- shareholder

- environment
- technology
- time/speed

Stage 2 – the offering
This stage represents the type of responsiveness an organization exhibits when operating in the marketplace. Benchmarking is a very powerful tool in building a strong capability of delivering goods and services. Capability represents the various levels of energy, strengths and competences that any organization should be able to control and harness effectively. The pull forces include (Fig. 1.2):

- creativity and innovation
- teamwork
- streamlined processes
- technology
- measurement
- a culture of continuous improvement

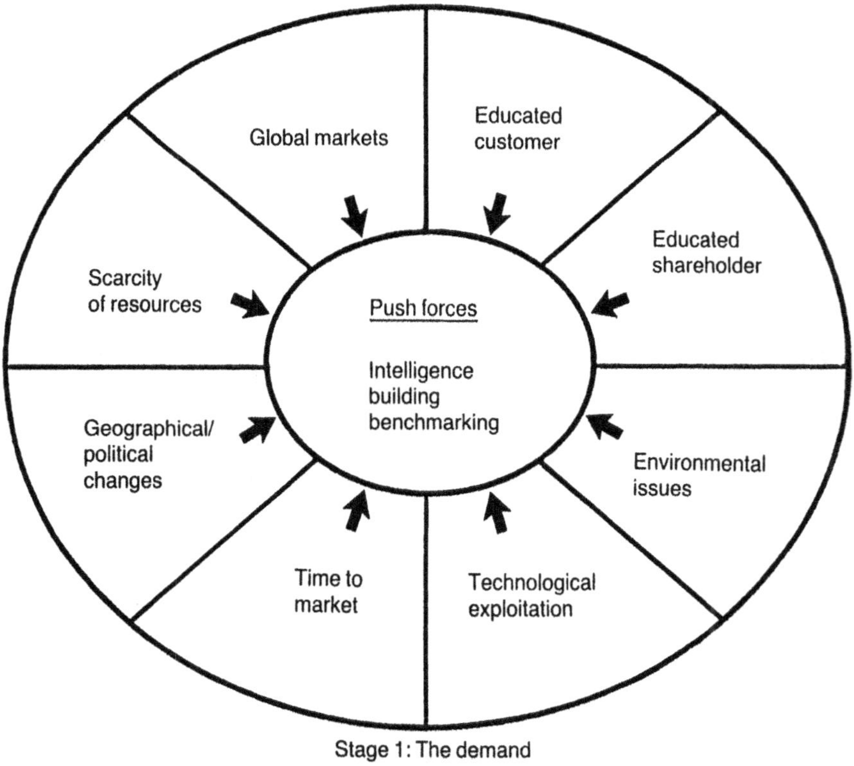

Figure 1.1 A model of establishing modern competitiveness.

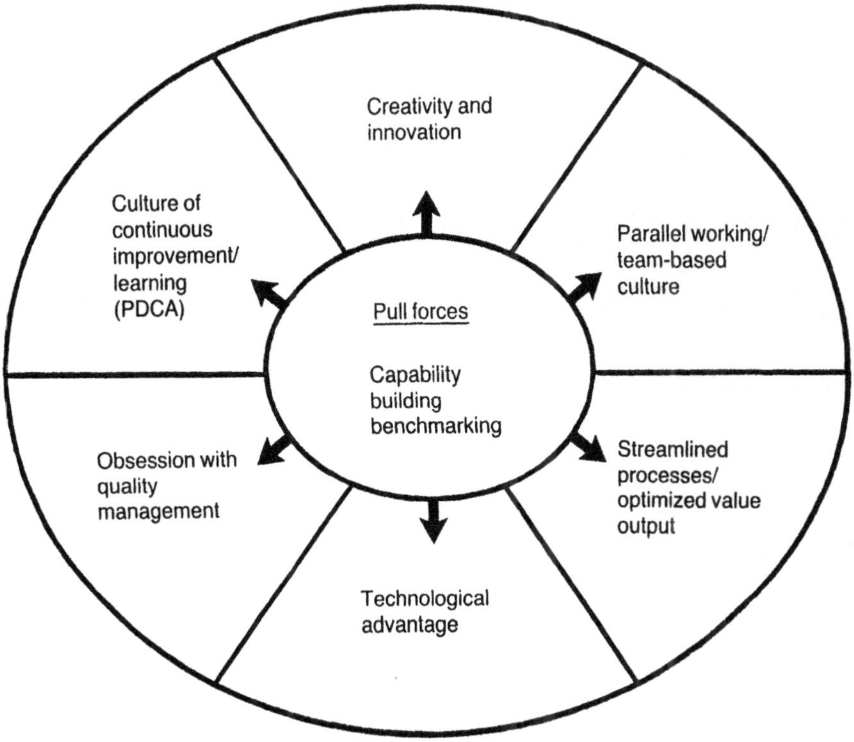

Figure 1.2 A model of establishing modern competitiveness.

This model can be described in terms similar to a living cell. Cells live, grow and prosper according to their ability to protect themselves from adverse bodies and their ability to create a good climate and nourish themselves to become strong and healthy.

2

The total quality movement: where did it start, what does it mean, and how does it affect business organizations?

This chapter is intended to trace the evolution of TQM, its use in Japan, how companies in the West have pioneered it (with specific examples from companies such as Rank Xerox) and what the benefits are.

2.1 Background

By the early 1970s, traditional markets in the West were being undercut by foreign competitors able to take advantage of lower labour and overheads, strong Western currencies and a world oil crisis.

The initial response from the West was to reduce costs further, then to drop low-profit items from their production lines, expand production operations in Third World countries and purchase increasing numbers of foreign parts and components.

Gradually Western consumers became aware that the cheaper imported goods, especially those from the Pacific basin, including Japan, Taiwan and Hong Kong were also associated with better quality and they began the move away from traditional home produced products and services to those of new, quality conscious, foreign competitors.

Many industries were soon devastated by this competition: motorcycles, cotton and textiles, consumer electronics, computer technology, motor cars – and the list goes on.

Competition had shifted from the basis of price alone to price and quality.

When Western companies discovered that company-wide quality control had helped give Japanese companies a competitive edge they began to try to do something about it.

In the beginning some Western companies adopted the 'quality circle' approach. Employees were trained to use basic statistical tools such as Pareto charts, cause and effect diagrams, histograms and control charts.

Small groups, called quality circles or employee involvement groups, were formed and, with the help of trained facilitators, were helped and encouraged to resolve their own business problems. As an example, Rank Xerox (UK) Ltd in 1983 set up an employee involvement pilot scheme. During 1984, some 15 groups were formed and by mid 1986 in excess of 80 groups were active.

Some companies tried to involve middle managers in quality development programmes, whilst many still believed that traditional quality assurance or quality control techniques would resolve their quality problems.

As recently as 1989, a survey conducted by the American Electronics Association revealed that although 85% of respondents had undertaken a quality improvement effort, fewer than one third could document significant improvements in quality and productivity.

Too many quality programmes in the past had been narrow; quality control efforts were concerned with catching mistakes and less than acceptable product after production, rather than with turning out high-quality products the first time around and improving the quality of every activity of the entire organization.

Successful companies such as 3M, Ford, Hewlett-Packard, Rank Xerox and ICL began to realize that quality could not be inspected into a product.

No single organizational action, like quality circles, Taguchi methods, investment in technology, etc, can dramatically improve quality.

Quality is defined by the customer and a great deal of quality problems originate in service or administration areas. Therefore, delivery of the right quality every time requires the involvement of everyone who works in the organization. Customer quality has to be an integral part of every practice, policy, process and procedure.

In the West, this approach is known as total quality management (TQM).

The success of international competitors who take quality seriously, coupled with the rising expectations of customers suggest that companies who intend to prosper have no option but to incorporate quality as a major part of their strategic approach.

It is impossible to simply copy the Japanese approach because of the substantial differences in culture and background. However, the concept of TQM is beginning to capture the imagination of chief executives and senior management, not only in manufacturing but also in service industries, health care and educational institutions. Governments are developing national quality campaigns and business schools and universities are becoming more committed to quality management training.

TQM incorporates the teachings of early pioneers such as Deming, Juran, Feigenbaum and Ishikawa and is still evolving with present day disciples such as John Oakland and Tom Peters. Its major premise is that quality is the key to business success in the 1990s and thereafter, and that this, rather than price or delivery is the key to competitive advantage. There are 'hard' and 'soft' aspects to TQM. The former are concerned with production techniques, including statistical process control (SPC), just-in-time inventory control (JIT), the seven basic statistical tools and policy deployment (*hoshin kanri* or *hoshin* planning). The soft element of TQM is concerned with creating awareness of the need to focus on meeting the needs of internal and external customers and has a great emphasis on the management of human resources.

2.2 The evolution of TQM

Before we look in any detail at the evolution of TQM let us consider some meanings and definitions.

So, what is quality? Is it a state, a condition, a feeling, or an impression? Does quality have any components and, if so, what are they?

Definitions of the word quality can vary enormously. For example, at one extreme it can be considered as:

- that which makes a being or thing such as it is – a distinguishing element or characteristic;
- the characteristics of anything regarded as determining its value, place, worth, position, etc.;
- quality is neither mind nor matter, but a third entity independent of the two – even though it cannot be defined, you know what it is.

At the other:

- quality is the degree of excellence at an acceptable price and the control of variability at an acceptable cost.

Quality refers to certain standards and the ways and means by which those standards are achieved, maintained and improved upon. Most definitions given to quality refer to 'fitness for purpose' or 'conformance to requirements'.

The standard definitions on quality have been given by various institutions such as the British Standards Institution (BSI), the American Society for Quality Control (ASQC), the European Organization for Quality Control (EOQC) and the International Organization for Standardization (ISO) amongst others.

Quality therefore is the totality of features and characteristics of a product or service that bears on its ability to satisfy given needs (BS 4778(1);ASQC A3 1978).

The characteristics of quality are given below.

- **Grade** A category or rank indicator of products, processes or services intended for the same functional use, but with a different set of needs (ISO/TC 176(1984)).
- **Imperfection** A departure of a quality characteristic from its intended level or state without any association with conformance to specification requirements or to usability of a product or service (ASQC A1 (1978)).
- **Nonconformity** A departure of a quality characteristic from its intended level or state that occurs with a severity sufficient to cause an associated product or service not to meet a specification requirement (ANSI/ASQC A1 (1978)).
- **Defect** A departure of a quality characteristic from its intended level or state that occurs with a severity sufficient to cause an associated product or service not to satisfy intended normal, or reasonably foreseeable, usage requirements (ANSI/ASQC A1 (1978)).

Gradually, quality became to be more associated with what the customer wanted and thus definitions began to appear with this point of view in mind. For example:

- Quality is driven by the marketplace, by the competition and especially by the customer.
- Quality is a key attribute that customers use to evaluate products or services.

So, from the customer's point of view, what characteristics of a product or service determine its quality? A few of these are listed below. Firstly, from an objective standpoint:

- How well does the product perform against expectations?
- How well do the features (the secondary characteristics which supplement the product's basic functioning) meet expectations?
- How reliable is the product?
- How well do the product design and operating characteristics conform against pre-established standards?
- How durable is the product, from both economic and technical standpoints?
- How well is the product serviced? (Speed and competency of repair.)

Secondly, from a more subjective standpoint:

- How well is the product serviced? (Courtesy of telephone contact, repair staff.)
- How well does the product meet the aesthetic needs of the customer? (How does it feel, sound, look, taste, smell?)
- How does the customer perceive the quality of the product when comparing product brands?

TQM evolved from this growing need to understand and consistently meet the rising expectations and requirements of the customer. The main precursors to TQM were quality control (QC), total quality control (TQC) and quality assurance (QA), as illustrated in Fig. 2.1.

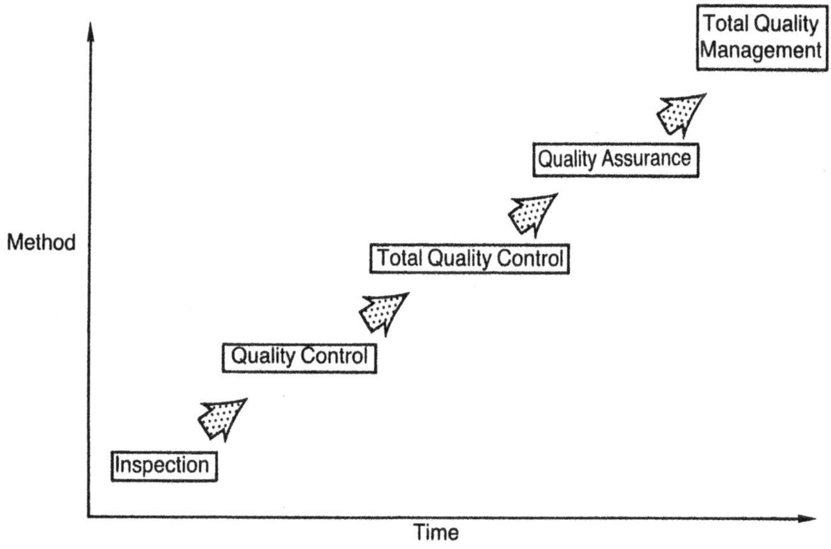

Figure 2.1 The evolution of TQM.

Quality control (QC) is defined as follows:

- Operational techniques and activities aimed both at monitoring a process and eliminating causes of unsatisfactory performance of relevant stages of the quality loop (quality spiral) in order to result in economic effectiveness (ISO 8402 1986).
- The operational technique and the activities which sustain a quality of product or service that will satisfy given needs (ANSI/ASQC A3 1977).

QC is therefore the use of techniques, mainly statistical, to achieve, maintain and try to improve on quality standards of products and services. It co-ordinates the links between:

- specification of what is required;
- designing the products/services required;
- production/installation/assembly of parts, components, elements of the service/product package;
- inspection of the service/product package to determine conformance to customer specifications;
- monitoring usage/consumption of product/service to feed back information for improvements wherever possible.

The principles of QC are based on four 'No's.

- No manufacture without measurement.
- No measurement without records.
- No records without analysis.
- No analysis without feedback and corrective action.

The concept of control includes all the activities which enable organizations to achieve their objectives efficiently and economically. It is an ongoing process based on continually trying to rectify and improve. Basically there are three types of control:

- **irregular** spasmodic approach to controlling quality, e.g. when a customer complains;
- **routine** regular inspection at specific stages of the process which is providing the product/service; and
- **scientific** control through measurement and analysis through statistical sampling.

Statistical Quality Control (SQC) measures the degree of conformance of raw materials, processes and products to previously agreed specifications. It uses control charts for measuring variations in, for example, weight and/or dimensions and for establishing the range between the smallest and largest readings.

Total Quality Control (TQC) has been defined in a variety of ways, all of which emphasize its important role in focusing on the various activities within organizations. It has, for example, been described as:

- A management framework to ensure continuing excellence.

Within this framework is contained the basic philosophy of TQC.

- It is a business philosophy which links together all the functions of the organization e.g. sales, marketing, manufacturing, etc., by two-way flows of information.
- It has a mind set which only approves criteria which will lead, via the use of continuous improvement, to better than acceptable quality.
- It is a process for continuous improvement where current standards constantly present opportunities for the achievement of new, higher targets.
- It provides reliability and consistency in the delivered product/service.

Some other definitions of TQC are given here.

- The application of a number of important management principles and the statistical control of quality to all stages of planning, design, production, service, marketing, accounting and administration. It aims at achieving disruption-free, error-free activities that produce defect-free products and services at a quality cost suited to their market.

- An effective system for integrating the quality development, quality maintenance and quality improvement efforts of the various groups in an organization so as to enable marketing, engineering, production and service at the most economical levels which allow for full customer satisfaction.
- Problem analysis in order to develop long-term solutions rather than response to short-term variations. Concern with direct cost reduction and a preoccupation with efficiency are ousted in favour of the pursuit of quality through the elimination of waste and non value-adding procedures and assuring continuous improvement through the refinement and expansion of quality control systems and procedures.

Company-wide Quality Control (CWQC) is the name used in Japan for TQC. The emphasis here is once again on the total control of quality across the entire organization and with the explicit contribution of every department and every individual. CWQC looks at the process of serving customers from the point of view of the customer–supplier chain (internally and externally). It therefore demands that individuals' roles are conducted with alacrity, quality and efficiency, etc.

Quality Assurance (QA) means basically that QC is conducted in a systematic manner.

QC means checking that various stages of the process of serving the customer have been conducted correctly and any defects identified have been corrected.

QA means that the process of checking, correcting and controlling is conducted in such a manner that the manufacturer/service provider is aware that all stages of the process are being conducted correctly (with the set quality standards in operation) and that what is planned is what is expected to result in terms of output. QA also means there is a set of documentation (a system) which demonstrates the existing standards of quality and reliability. QA uses what has been referred to as the 'death certificate approach'. It rejects inspection as the answer to quality problems and encourages the implementation of procedures in order to comply with set standards. QA has been defined as follows:

> Quality Assurance contains all those planned and systemic actions required to provide adequate confidence that a product or service will satisfy given requirements for quality (ISO 8402 1986; ANSI/ASQC A3 1977).

To ensure that products and services are in compliance to set standards, QA relies on the use of SPC techniques. The system for the implementation, controlling and auditing in QA is often open to third-party approval either by customers or government agencies.

Total Quality Management: how then does this differ? TQM is the management of total quality. Quality management is defined as:

> That aspect of the overall management function that determines and implements the quality policy (ISO/TC 176 1984), and as such, is the responsibility of top management (ISO 8402 1986).

Since it has been established that quality management is a managerial responsibility, one therefore has to link it to the destiny of businesses. It is not just a question of achieving standards but one of survival and being strong all the time. Furthermore, managerial responsibilities are not just concerned with focusing on one particular aspect of the business but in being fully aware and in control of all the various activities no matter how small they are. This leads to the argument that TQM is an organizational concern and not the domain of any specialists or specific functions.

Leadership is perhaps the most important ingredient in the philosophy of TQM and has been addressed in detail by early pioneers such as Deming, Juran and Crosby.

A ship's captain has the responsibility of keeping his ship afloat, on course, guaranteeing its safety and ensuring the safety of the crew until the planned destination has been reached. This analogy applies equally well to the leadership of a company. Leadership is often accompanied by a series of actions and initiatives which lead to positive outcomes, reflecting organizations' ambition and desire to succeed.

Within the framework of TQM the role of the leader will include:

- the provision of unquestioned leadership;
- an explicit focus on the customer;
- the training of all employees;
- the recognition and reward of employee participation;
- communication about quality both internally and externally;
- the provision of a quality process and quality tools.

The change model used by Rank Xerox over the past ten years supports this point (Fig. 2.2).

Another analogy would be likening the role of a business leader in implementing a TQM philosophy to that of a general preparing for a long, hard battle. Von Clauswitz commented:

> Most battles are won, or lost, before they are engaged, by men who take no part in them; by their strategists.

It is now widely accepted that quality is also a behavioural question and that one of the major tasks for senior managers is to change people's attitudes and understanding of the meaning and importance of quality. Most of the published literature seems to be concerned with broader issues such as the implementation of TQM, the utilization of useful statistical tools and techniques such as SPC and the benefits of encouraging a problem-solving approach to doing business. Research, however, points

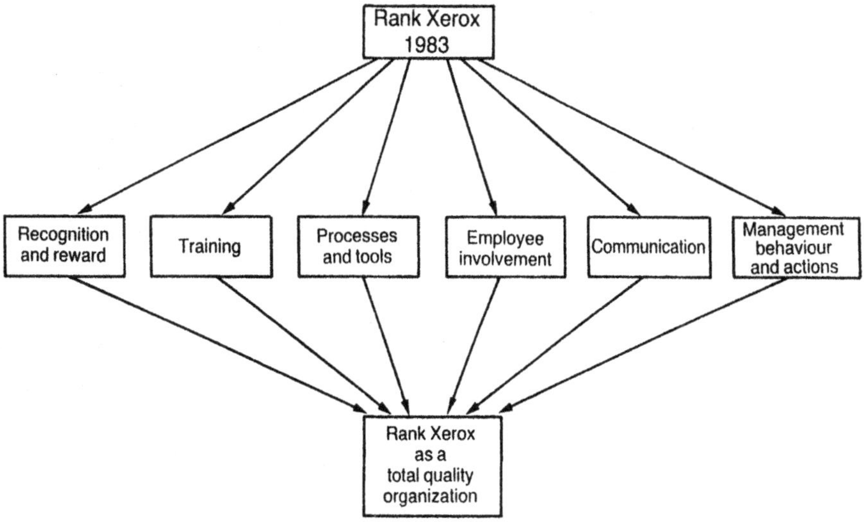

Figure 2.2 The Rank Xerox Hi-level change strategy.

alarmingly towards the total ignorance of understanding of what TQM is really all about at the various levels of business organizations.

It is therefore not just about achieving certain standards of competitiveness or introducing new techniques, concepts, methodologies and technologies; it is about changing the attitudes and behaviours towards doing business whereby parameters of competitiveness are set by, or negotiated with, the customer. One conclusion which may be drawn from this research is that managing quality is the job of each and every member of an organization, not solely of its managers. In this sense, managers as well as shopfloor workers must shape the environment which is most propitious for such an achievement and be given incentives to ensure that it becomes, in the routine tasks of everyday life, a reality.

TQM has been defined as a philosophy based on the quest for progress and improvement. In this sense TQM looks for continual improvement in the areas of cost, reliability, quality, innovation, efficiency and business effectiveness. TQM is certainly not an outcome of blueprints and simulated business performance. It is a dynamic way to perform with a determination to improve in all areas.

TQM is a fundamental shift from what has gone before. The systematic analysis, pre-planning and blueprinting of operations remains essential, but the focus switches from a process driven by external controls through procedure compliance and enhancement to a process of habitual improvement, where control is embedded within, and is driven by, the culture of the organization. It is an approach for continuously improving the quality of goods and services delivered through the participation of all levels and functions of the organization.

The Rank Xerox quality policy accurately reflects this:

> Rank Xerox is a quality company. Quality is the basic business principle for Rank Xerox. Quality means providing our external and internal customers with innovative products and services which fully meet their requirements. Quality improvement is the job of every Rank Xerox employee.

TQM looks for continual improvement in the areas of cost, reliability, quality, innovation, efficiency and business effectiveness. It is rarely, if ever, possible to achieve this within the boundaries of one organization. Benchmarking is, therefore, a critical tool of TQM which, when properly used, ensures that no stone is left unturned in the search for, and integration of, best practices, wherever they are to be found.

3

Origins of benchmarking and its meaning

This chapter will examine the history and development of benchmarking and will examine the meaning and definitions.

3.1 Introduction

Levelling is the operation, carried out by the Ordnance Survey (O.S. Leaflet 33 dated May 1977 and October 1983), of measuring differences in height between established points, relative to a datum. Heights are identified by means of **bench marks** placed at many locations throughout the country.

A bench mark is a mark whose height, relative to ordnance datum, has been determined by levelling. The types of bench marks used by ordnance survey are:

- Fundamental bench marks
  These marks are constructed at specially selected sites where foundations can be set on stable strata such as bedrock or **benches** (Bench n. *Geog.* 'A terrace formed in rocks; also, elevated ground along the bank of a lake or river'). In this way, the likelihood of movement of the mark is minimized. They provide a system of stable marks throughout the country which controls the whole levelling network. They are sited at intervals of approximately 40 km along lines of geodetic levelling.
- Flush brackets
  These consist of metal plates about 90 mm wide and 175 mm long cemented into the face of buildings. They are fixed, where possible, at intervals of about 1½ km along all the lines of geodetic levelling. A flush bracket is also set into one of the sides of most Ordnance Survey triangulation pillars.
- Cut bench marks
  These are the commonest form of bench mark. They consist of a horizontal bar cut into vertical stonework, brickwork and concrete surfaces. A broad arrow is cut immediately below the centre of the horizontal bar.

There are several other types of bench marks, which, together with the ones described, form a comprehensive network. Bench marks are provided to meet normal user requirements at densities which will vary according to the nature of the area. The network is designed so that the maximum distance to the nearest bench mark will be reasonably consistent in similar types of terrain and for similar degrees of development of the land.

Technology has established that the precision with which individual bench marks have been located and the accuracy of the bench mark network itself is second to none.

What, therefore, is benchmarking (one word)?

We have established that a bench mark is an extremely accurate device and is therefore a standard of excellence worthy of emulation. 'Benchmarking' is a word which appeared, as it were, over the Total Quality horizon two or so years ago and many seminars and workshops are being organized and many articles and books are being written about the subject of 'benchmarking'.

But what is it?

Going back to the Funk and Wagnell standard dictionary international edition of 1960 there is no reference to benchmarking. If, however, we move forward to the 1987 edition of the Oxford Reference dictionary, although we still find no reference to benchmarking, we do find the following reference:

> Benchmark – *A surveyor's mark indicating a point in a line of levels; a standard or point of reference.*

Is this 'standard' the beginning of today's use of the word benchmark? – a standard upon which we should all set our sights? We believe so.

Further evidence to support this view is to be found in the latest edition of Roget's Thesaurus which links the word benchmark with words and phrases like standard of comparison, yardstick, frame of reference, and model. So, we begin to see the origins of the word benchmark as it is used today. The term benchmarking has become synonymous with the means of identifying the benchmark.

3.2 Why is benchmarking so important?

In the year 500BC, Sun Tzu, a Chinese general, wrote:

> 'If you know your enemy and know yourself, you need not fear the result of a hundred battles.' (Sun Tzu, *The Art of War*).

Sun Tzu's words could just as well show the way to success in all kinds of business situations. Solving ordinary business problems, conducting management battles, and surviving in the marketplace are all forms of war, fought by the same rules – Sun Tzu's rules.

Another truth is a simple word of unknown age. It is the Japanese word *dantotsu*, meaning striving to be 'the best of the best'. It is the very essence of benchmarking. We in the West have no such word, perhaps because we have always assumed we were the best. World competitive events have smashed that notion forever (Camp, 1993).

In the late 1970s Xerox, as profits and market share tumbled, began to realize that Japanese photocopying companies were selling their products cheaper than Xerox could manufacture. Xerox began to compare itself directly with this new competition in terms of unit manufacturing costs, manufacturing methodologies, time to market and so on, to understand what had to be done in order to stay in business.

Although it is fairly well established now that the Xerox Corporation was the pioneer of benchmarking, it is only fair to say that in 1979 as an organization with a virtual monopoly in its sector, Xerox had become overly bureaucratic and a little complacent. By 1980, it became apparent that the company had to improve productivity rapidly to compete effectively in the global market.

Japan already dominated the video and motorcycle industries. Japan's success in microwave ovens, sports goods and cars was still on the horizon. It was evident that one of their next target markets was copiers. This was not a market share problem; this was an issue of survival. An in-depth study took place which recognized the need to overhaul the company. The company began to evaluate itself externally through this process which became known as competitive benchmarking.

The results from benchmarking were startling:

- Xerox's ratio of indirect to direct staff was twice that of direct competition;
- it had nine times the number of production suppliers;
- assembly line rejects were in the order of ten times worse;
- product time to market - twice as long; and
- defects per 100 machines - seven times worse.

Unit manufacturing cost was the same as competitors' selling price. The assumption prior to benchmarking was that the competitors' machines were poor quality. This was proved to be wrong, and to drive the point home their competitors were making a profit!

The company was vulnerable. In every significant measure of performance, Xerox exceeded the benchmark levels set by the Japanese. The company had nine times more suppliers, was rejecting ten times as many machines on the production line, and taking twice as long for their products to reach the market.

By contrast, benchmarking revealed the Japanese joint venture Fuji Xerox was performing well, attributing its success to quality principles and practices. These competitive benchmarking findings brought home the reality of the business crisis and forced senior management to acknowledge both the size of the problem and the nature of change required to revitalize the company.

The strategic plan at that time looked to 8% productivity growth year on year, compared to an industry average of 3%. Benchmarking revealed that this would never catch the Japanese. An 18% growth, year on year, would take five years to catch up. The strategy selected as a result of this was a worldwide commitment to total quality. Internally this strategy is known as **leadership through quality**, i.e. market leadership through the application of total quality. It has proved to be not only the salvation, but also the foundation of the company today and what it will be tomorrow.

The benchmarking exercise, which involved senior management as well as functional experts across the entire company, represented a new, externally focused way of looking at the world. Indeed it has been acknowledged as one of the first concrete activities that got the company on its feet and moving again. To quote the *Financial Times*:

> The power of benchmarking is underlined in the case of Xerox with it being one of the main factors behind the company's revival in the 1980s.

3.3 The evolution of benchmarking

For Rank Xerox, benchmarking was an effort to take conventional competitive analysis one step further. Its roots began in 1979 in the manufacturing operations when it encompassed an in-depth, ongoing study of best competitors, a detailed reverse engineering of competitor products, technology processes, what they achieved and how they did it. Operating capabilities and features of competing products were subjected to a tear-down analysis.

As experience was gained, success achieved and the company learned more about itself and the best practices in industry, the need to change the total company became increasingly clear. The focus was extended beyond manufacturing and the establishment of benchmarks and goals, to an emphasis on the processes used by the best in class as well as the competition. From an activity in relative isolation, benchmarking was extended as a strategic quality tool to all aspects of the business and progressively integrated into the management process.

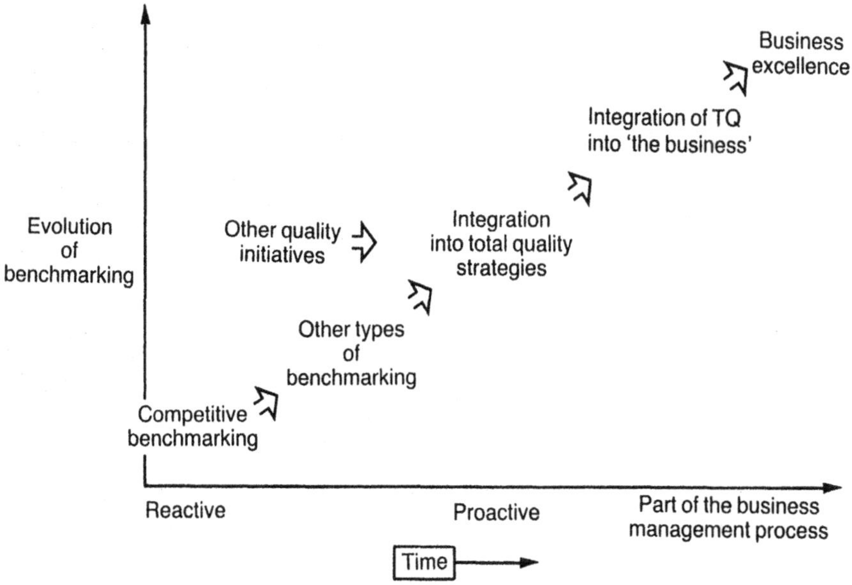

Figure 3.1 The evolution of benchmarking.

3.4 Benchmarking today

Benchmarking is the continuous process of measuring products, services and processes against the strongest competitors or those renowned as world leaders in their field.

The working definition is the search for industry best practices that lead to superior performance. Benchmarking is used at the strategic level to determine the standards for performance against four corporate priorities,

- customer satisfaction
- employee motivation and satisfaction
- market share
- return on assets

and at the operational level to understand the best practices or processes that help others achieve world-class performance.

These inputs help to set realistic targets in the business plans (because it is known the standard *is* achievable), and to identify the specific actions and resources required to improve performance. The aim is not only to match existing performance levels – but to exceed them and to become world-class in areas that provide competitive advantage. The process is continuous because of the high rate of change in the business environment and consequently benchmarks are continually redefined.

In summary, benchmarking is very much an opportunity for an organization to learn from the experience of others. As with all learning experiences, there are factors conducive to success or failure. For example, complacency or high security on the part of an organization does not lend itself to a successful benchmarking exercise. On the other hand, a shock experience can stimulate the self questioning and self doubt which is often the necessary prelude to growth. A constructed crisis and challenging of ideas is, *per se*, a beneficial experience. However, to translate a beneficial educational experience into concrete actions, a structure and process has to be in place.

Unless there is an appreciation of the sources of resistance to change in an organization and involvement of key players in a benchmarking exercise, then efforts will be to no avail.

There are many examples of technically outstanding benchmarking exercises which have failed to grasp this fundamental prerequisite.

By the same token, some of the more superficial benchmarking events, when senior managers have visited other organizations, can work because they have provided a first hand vision and tangible examples of what can be achieved. They can provide a motivational source of energy. However, unless there is a follow-up mechanism in place and reinforcement through a systematic process, then the learning from such visits can fade. It is also true to say that accounts and enthusiasm from benchmarking visits can be as interesting to others as a colleague's holiday snaps. There is nothing like a 'prophet in a foreign land' syndrome to incur antagonism and defensive reactions to the implementation of benchmarking findings.

To conclude, benchmarking is not a one-off analysis. It provides insights into a business and challenges its usual methodology. Implementation of the results and findings depends on a willingness to change and to adapt to new ways of doing things.

4

Benchmarking – an all encompassing tool

4.1 Introduction

Many companies today believe they are using benchmarking as an integral part of their improvement strategy. In reality, however, this is frequently not the case. The way benchmarking is being used in those companies is not driving for superiority, for stretched goals and reaching levels of excellence. They are only using it for incremental improvements, for cosmetic changes, for nudging and pushing the processes and practices in the right direction.

However, the best way to drive business results is through the management of business and work processes. In this context, benchmarking is a powerful, all encompassing tool which, if used properly, will assist enormously in driving the business forward. The following are several examples of where benchmarking can and should be used.

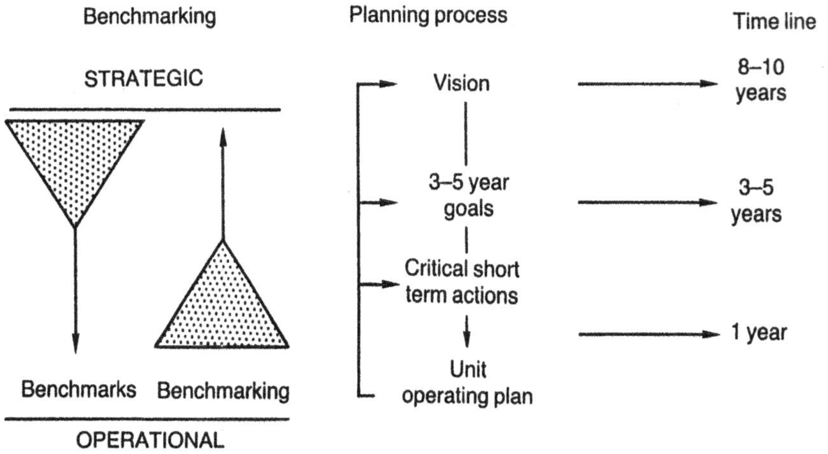

Figure 4.1 Benchmarks, benchmarking and the business planning process.

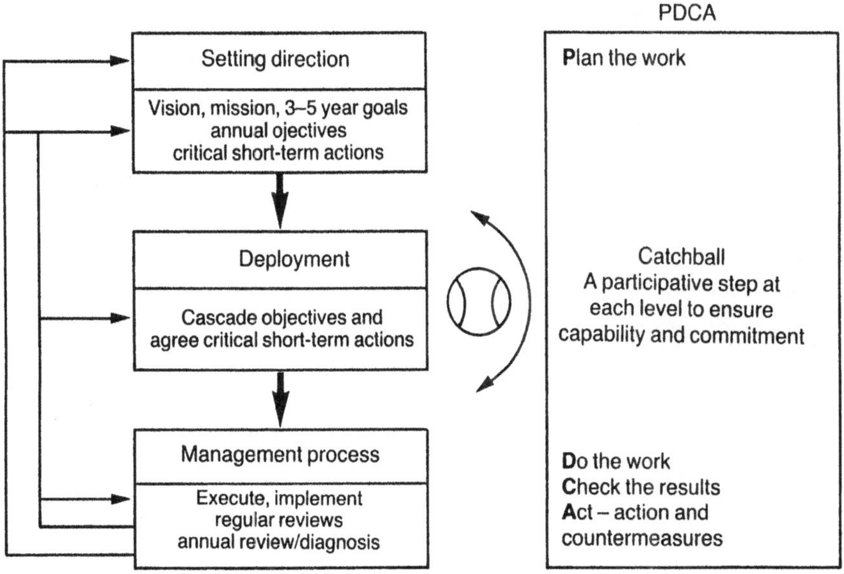

Figure 4.2 Policy deployment – process flow chart.

4.2 The importance of benchmarking

Benchmarking is important in the following areas.

- Business planning:
  - establishing a company vision;
  - setting of 3–5 year (strategic) goals;
  - identifying critical short-term actions for the achievement of the 3–5 year goals; and
  - putting together the unit operating plan.
- Policy deployment

  Setting of goals and targets is not enough. They have to be communicated to the organization so that everyone understands how his/her activities relate to the achievement of the critical short-term actions.

  Communication begins at the very top and cascades down. The senior management team first discusses its goals and strategies with its direct reportees to ensure understanding and acceptance. These discussions will include resource requirements, process capability, implementation strategy, etc., and continue until both parties are agreed on the goals and the means to achieve them. In order to reach agreement, it may be necessary to modify goals or adjust resource allocations.

  Once agreement has been reached, the communication/negotiation process is repeated down into the organization.

  Benchmarking has an important part to play in helping find the best practices which will assist in the achievement of the goals.

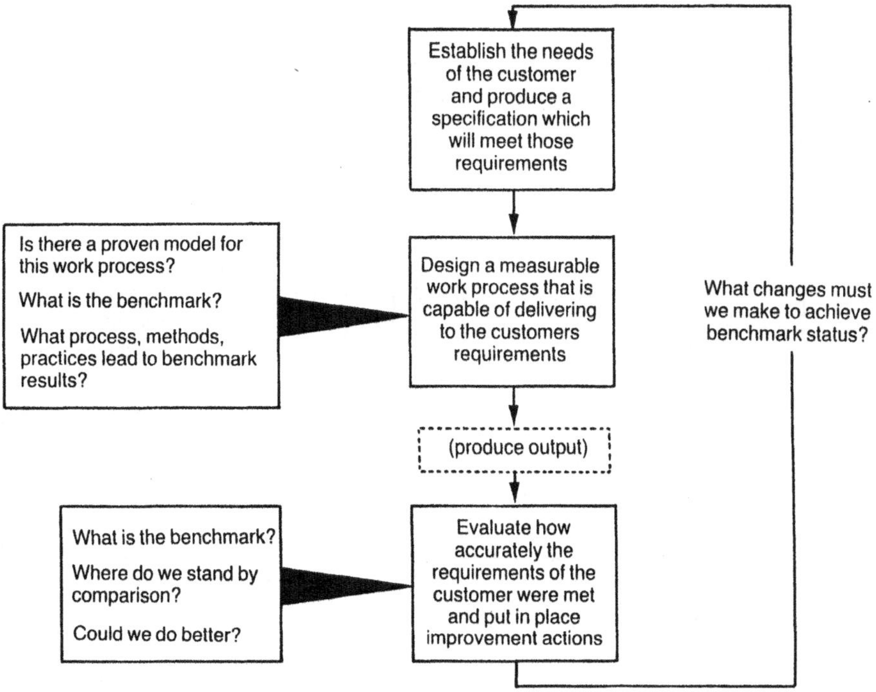

Figure 4.3 Benchmarking and continuous improvement.

- Solving business problems
  An important element of a total quality initiative is the ability to resolve business problems as quickly as possible. Many organizations employ a disciplined approach to this, using something similar to the model shown in Fig. 4.3. Benchmarking can be used to shorten the problem-solving cycle by answering questions similar to those on the diagram.
- Continuous improvement
  Another important element of a total quality drive is the need for continuous improvement. Many organizations have a formal process to enable this, perhaps similar to the one given in Fig. 4.3.
  Benchmarking is useful here in speeding the design of the work process and to understand how much improvement is possible.
- Customer satisfaction
  Customer satisfaction is fast becoming the number one goal of many companies today. Benchmarks can be established for key measures of customer satisfaction such as:
  - percentage satisfied
  - recommendations to others
  - repurchase intention
  - overall satisfaction

and to identify areas and levels of dissatisfaction. If the organization understands which business processes are behind the areas of dissatisfaction then benchmarking, along with problem solving, will eliminate them quickly.

- Cultural change
  The use of benchmarking over a period of time will gradually have an effect on the culture of an organization. As the focus for improvement becomes more and more based on external data, people will begin to challenge the status quo – be it a structure, a practice, a management system, or whatever.
  Questions arise, like
  - Am I doing the right thing?
  - Is there a better way of doing it?
  - Could I change certain things?
  - Suppose we did this . . . ?
  - What if . . . ?
  - Why are others doing it differently?

  while phrases like
  - We've always done it this way.
  - What was good enough for our founder is good enough for us.

  will no longer be acceptable.
- Continuous learning
  When everyone challenges the status quo, and people have the freedom to ask questions about what they are doing, why they are doing it in certain ways, and why they should change things, then and only then is there a learning organization – an organization constantly searching for better ideas, always curious, always hungry for innovation. This is continuous learning.
- Managing by fact
  Benchmarking brings facts to the decision-making process. It is effective in focusing the energy of the boardroom on the enemy – who is on the street, not across the table. So, benchmarking is helpful in taking out emotion and is therefore an effective cure for 'denial' problems. Most importantly, it helps the understanding of what needs to be changed and why.
- Business process improvement
  One of the major pre-requisites to carrying out a successful benchmarking study is a detailed understanding of one's own business processes. Once that is done, benchmarking helps to identify those processes which will achieve benchmark status with little more than the application of continuous improvement, those which require more rigorous simplification and those which are so far from the benchmark that re-engineering is required.

- Strategy development
  When benchmarking drives strategy development it gives day-by-day signals on how fast the market has moved since a strategic course was last set.
  The latest benchmarking data could, for example, be telling an organization:
  - we have moved too fast;
  - we need to re-trench; or
  - we need to change something.

  Once that way of thinking is in place, strategy can be flexible. Benchmarking drives the vision, the strategic goals aimed at delivering the vision, the short-term objectives that have to be achieved in order to make the vision become a reality, whilst constantly checking that the vision itself is sound.

The ultimate purpose of benchmarking is to help meet customer requirements. Aiming for, and meeting, customer requirements does several things. What is new for a customer delights for now, but it also raises expectations which forces targets to go up.

So in order to survive, you have to do a little bit more – you have to win the customer's loyalty on aspects of service, conditions of sale, professionalism, the brand name, etc., and so the only thing to do is to tighten the targets – beat the competitors in all aspects, on quality, on superiority of the product, on the excellence of service and also on efficiency and effectiveness.

4.3 The links between process and people

The links between process and people need to be fully understood. The work processes drive the business results. However, the people who work within the processes give a 'human value-added' contribution to the process productivity. Those people need to be constantly encouraged to keep looking for improvement opportunities in their bit of any particular process. This is the way the Japanese do it. Japanese employees know what the targets are, they are well educated, they are well informed on market conditions, they strive to meet those targets and they know it is going to be hard if they are not met.

This is true empowerment. They know that if the target is not met there is no bonus, because the results are poor. They know they have to do something internally to protect the organization. As customer expectations rise, people within the organization expect change on a regular basis. When this stage has been reached, the organization, be it a small department or a worldwide corporation, becomes much more flexible.

Benchmarking is the tool which allows this stage to be reached. It is the tool for educating, for communicating, for giving people the freedom to ask questions, empowering them, and building trust and loyalty.

Benchmarking is then an all encompassing tool, not just one used to improve a set of practices. It can drive at normal pace but also has enough power to drive much faster or even cause an organization to change tack or direction completely.

5

Benchmarking – the strategic tool

5.1 Customer – key determinant of strategic planning

As noted before, more companies are waking up to the fact that their success is greatly determined by the customer. Companies can no longer afford to develop objectives and measures which are internally driven. Markets are becoming more and more saturated, thus making growth and maintaining market share a real challenge. Customers determine parameters of competitiveness and therefore have to be listened to, and differentiation is no longer achieved through product/service excellence but through unique ways of delighting the customer. Strategies that lead to superior competitiveness are based on a clear capability to deal with changing customer demands and on the ability to respond quickly, effectively and economically. The design of the right strategies has to start with the customer. This is increasingly recognized by more and more organizations who in their mission and vision express the need to focus on the customer:

> Competitive realities are what you test possible strategies against. You define them in terms of customers. . . . Of course it is important to take the competition into account, but in making strategy that should not come first. First comes painstaking attention to the needs of customers.
>
> (Kenichi Ohmae, 1991)

> The challenge to every marketing professional is to justify his price premium to his customers.
>
> (David Pearson, MD, Sony Consumer Products, UK)

5.2 The link between competitive analysis and benchmarking

Competitive analysis is the most essential tool in strategic planning. Indeed, it is through a knowledge of competitors' strengths and weaknesses that effective strategies can be formulated. There are various avenues for analysing competitors and the market:

- competitor analysis using the product as the starting point (reverse engineering) to identify strengths, weaknesses, design capability and technology exploitation;
- financial analysis to give an indication of competitors' commitments, resource allocations, results, etc;
- environmental trends, to look at more issues; and
- market dynamics to understand consumer wants, attitudes, spending power, etc.

Competitive analysis is an excellent process for gathering data on competitors. It determines how big gaps are in terms of cost, quality of products and services, and timeliness of delivery in terms of new product launch, product to market cycle time, etc. As such, the information generated is essential for the formulation of strategies.

Competitor analysis does not necessarily lead to a good understanding of competitor behaviour. In fact, it is argued that competitive analysis does not really offer the necessary catalyst for senior managers to act confidently on closing the gap. Whilst the how much and how big have been answered, the what, why, how and when cannot be answered by competitive analysis. This is where benchmarking becomes essential.

Benchmarking is geared towards understanding the processes which lead to gaps in performance. It generates a high level of understanding of various practices associated with superior performance; thus it provides managers with realistic goals and gives them the confidence to develop strategies likely to lead to closing the gap.

5.3 Benchmarking strategic quality planning

The introduction of total quality management is often based on the need to translate customer requirements into products and services which will satisfy them in the short term and offer additional features to delight them in the long term.

Strategic planning and implementation must therefore be focused on waste elimination, productivity improvement and enhancing innovation and creativity for the benefit of the end customer.

Modern strategic planning processes can translate quality principles and their application at all levels. Frameworks such as the Malcolm Baldrige National Quality Award (MBNQA), the Deming Prize, and the European Quality Award (EQA) can be used to measure the effectiveness of strategic quality planning.

The MBNQA framework

Strategic quality planning is one of the seven elements of the MBNQA framework. It examines companies' planning processes and how quality is integrated into overall business planning. It covers:

- strategic quality and company performance planning processes (development of strategies, goals to lead to customer satisfaction, plans that address overall operational performance improvement, how plans are deployed, how evaluation is being carried out);
- operational quality and performance plans for the short term and the long term.

The Deming Prize framework
Company policy and planning is one of the ten criteria used for assessment under this framework. It assesses the policy for management, how quality control is determined and transmitted throughout the organization and how results are being achieved. In addition, the contents and appropriateness of the policy in question, its clarity and presentation are also scrutinized.

The EQA framework
The EQA framework has nine criteria covering the enablers for deploying quality. It also examines the ways in which performance measurement is achieved and results quantified. All of the criteria are related to policy and strategy and the organization's mission, values, vision and strategic direction are all assessed, as are the ways in which the organization sets about achieving them. The following are covered:

- how policy and strategy are based on the concept of total quality;
- how policy and strategy are formed on the basis of information that is relevant to total quality;
- how policy and strategy are the basis of business plans;
- how policy and strategy are communicated; and
- how policy and strategy are regularly reviewed and improved.

These three powerful frameworks therefore provide means of assessing the effectiveness of strategic deployment and how benchmarking can be applied to drive strategic thinking. The following examples reflect best practice in strategic quality deployment and the companies covered are all winners of prestigious quality awards. They have pioneered concepts of benchmarking, performance measurement, quality management, focus on customer satisfaction, commitment to process improvements and a vision of a creating/learning organization.

Strategic quality planning at Zytec Corporation
This company was the winner of the Malcolm Baldrige Quality Award in 1991. Zytec designs and manufactures electronic power supplies and repairs power supplies and CRT monitors. Its customer base include original equipment manufacturers. Zytec started to incorporate quality principles in January 1984. It chose three words as the company's new focus – quality, service and value.

The planning process at Zytec uses quality to deliver to the end customers. This process is heavily based on data collection, which is obtained through soliciting customer feedback, conducting market research and benchmarking customers, suppliers, competitors and industry leaders. Goals are then set by long-range strategic planning using a cross-functional team approach and a focus on core processes.

Zytec encourages the involvement of every employee in the development of the long-range strategic plan. The long-range strategic plan is also shared with major customers and suppliers who are asked for their initial reactions to it. Strategic planning is deployed through a process called **management by planning** (MBP). This process involves employees in the establishment of long-range plans and short-term objectives. Detailed action plans to implement the goals are put together by departmental teams and submitted to the top level.

To support the achievement of departmental goals, the short-term action plans are converted to financial plans. Each department has detailed budgets, and its manpower plans and capital plans have to be established to support the objectives. Regular reviews take place every month to assess progress.

Strategic quality planning at Cadillac Motor Car Company

Part of General Motors, North American Automotive Operations, the company was founded in 1902 by Henry Martin Leland. In 1987, it was re-organized into a single business unit to add engineering and manufacturing excellence to its marketing capability. It produces luxurious cars for the upper end of the market.

At Cadillac it is firmly believed that the business plan is the quality plan. The business plan is designed to ensure that Cadillac reaches the highest standards of quality and customer satisfaction. It aims to achieve three things:

- to ensure Cadillac's mission and strategic objectives are in line with both General Motors' corporate mission, and the business environment;
- to guide internal processes such as the use of simultaneous engineering and the establishment of quality networks; and
- to build discipline into the development and achievement of specific short-term and long-term quality improvement goals.

The business planning process at Cadillac begins at the senior level with a review of the mission's strategic objectives to ensure there is complete alignment with corporate group plans. A situation analysis is then performed. As a result of this, a draft of the proposed business objectives is produced with targets for the following year. This is communicated throughout the organization to gather feedback on its appropriateness and achievability.

The revised business plan is then shared with various unions at management levels throughout the company. Normally, the business objectives describe areas where the company wishes to achieve performance targets – quality competitiveness, disciplined planning and execution, leadership, people.

Deployment of the business objectives is then carried out. Teams are encouraged to translate the business objectives into performance standards at local levels. These are then reviewed by the senior management team, and once there is total agreement on the business plan and its deployment, wider communication takes place and the implementation phase starts.

The business planning process at Cadillac is heavily driven by benchmarking. At all levels people are encouraged to use competitive analysis to establish product and process benchmarks. For instance, product comparisons take place between Cadillac vehicles and those rated as best in class, and process comparisons are made against companies that demonstrate world-class performance, regardless of their product or service. Like Zytec, Cadillac is also committed to involving its partners. Key suppliers are included in **partners in excellence conferences.** Cadillac also encourages dealers to be involved in the planning process through the National Dealer Council, an advisory board to the executive staff.

Strategic quality planning at Wallace Co. Inc.

This company is a distributor of pipe valves and speciality products for the refining chemical and petrochemical industries. Wallace is also a winner of the Malcolm Baldrige National Quality Award. A senior management team develops the annual quality strategic objectives (QSOs) which drive the quality process at Wallace. Wallace's quality planning process is customer focused, and it relies heavily on data which is customer related. Sales associates closely monitor market analysis reports, and trends in customer complaints and changes in customer specifications or audit criteria are used to determine customer needs for the future.

The quality management steering committee (QMSC) analyses all the data at monthly meetings. Employees are encouraged to contribute at all levels to the planning process by participating in the annual revision of the quality business plan. In addition, quality task forces at all levels provides ideas for areas of improvement which they think should be added to the business plan. Suppliers are encouraged to contribute to the development of the business plan. Supplier quality is monitored and improvements are mutually developed through Wallace's vendor certification programme.

Once the planning of priorities has been carried out by the QMSC, implementation is carried out through quality improvement process teams empowered to take the necessary steps. The plans are then disseminated to all areas with clear responsibilities assigned.

Wallace uses benchmarking extensively in the development of its business plan. It relies on a wide variety of sources including the regional reputation of firms, customer input, literature references, process characteristics critical to Wallace and the customers, the involvement of the chemical and petrochemical industries which comprise nearly 85% of Wallace's customer base, quality practices of previous Baldrige winners as well as other firms in the distribution industry.

Strategic quality planning at Ritz–Carlton

The Ritz–Carlton, winners of the Baldrige Award in 1992, drives its business plan using quality principles. The primary objectives are to improve quality of all its products and services to reduce cycle time and improve price value and customer attention. It relies heavily on benchmarking studies both within and outside its industry, and it has developed a disciplined integrated planning system. It encourages employees and managers at all levels within its business and relies on teams given the specific task of setting objectives and devising action plans which are reviewed by the corporate steering committee. The task of the teams are to enhance quality and productivity of the Ritz–Carlton by, for example:

- aligning all levels around the Ritz's common vision and objectives;
- encouraging all of Ritz's people to think beyond the demand of day-to-day activities;
- increasing communications among the diverse functions that make up the Ritz–Carlton; and
- using simultaneous integrated problem solving.

Action plans are then put in place to implement the business objectives. For example, the top priorities for the Ritz–Carlton include becoming defect-free by 1996. The ultimate plan for the Ritz–Carlton is to become the first hospitality company with 100% customer attention and their ultimate goal is to use time, money, and people in order to achieve total customer satisfaction. This is carried out by managing processes through reductions of cycle time and streamlining activities to provide continuous price/value improvement to all the Ritz's customers.

Strategic quality planning at Texas Instruments Defence Systems and Electronics Group

Winners of the 1992 Malcolm Baldrige Quality Award, TIDSEG has grown from being a small navy contract company in 1942 to a $2B defence electronics company 50 years later. It has always focused on the end customer: 'Texas Instruments exists to create, make, and market useful products and services to satisfy the needs of its customer throughout the world.' This statement, issued in 1964 by a former TI Chairman, has been used throughout the years to guide strategic planning for the company for almost three decades.

Similar to previous examples, TI believes that its business planning process is synonymous with its total quality planning process. Their overall approach to strategic planning is characterized by a focus on:

- customer needs and expectations;
- specific products and technologies;
- continuous quality improvement;
- involvement of people at all levels;
- multi-customer driven long-term goals; and
- short-term plans that are derived from long-term goals.

Strategic planning at TI begins with a thorough understanding of customer needs. In addition to multi-level, multi-functional involvement for the development of its business plan, TI strongly believes that quality and strategy ought to be nothing but the same. It believes that this is the most effective way for integrating total quality at all levels and ensuring that customer satisfaction becomes the responsibilities of every single employee. TI relies on business process mapping (BPM) methods, and also benchmarking techniques in order to improve continuously its strategic planning process and to ensure that it focuses continuously on the end customer. The strategic planning process at TI enables it to define its short-term and long-term goals for quality improvements and the achievements of company performance standards. For example, the current short-term goals include:

- defect reduction by a factor of 6 every year;
- 25% reduction per year in cycle time;
- structuring the business for competitiveness – enhancing profit after tax/return on assets; and
- increasing net revenue per person.

Texas Instruments puts forward the following enablers in order to achieve the above goals:

- 6 Sigma training and deployment;
- continued deployment of SPC and design of experiments (DOE) using BPM and benchmarking of key processes;
- integrated product development process;
- organizational flattening to five levels;
- increased training for all employees;
- supplier-based reduction; and
- continued integration of the Baldrige Assessment Criteria.

In addition to the short-term goals, Texas Instruments has two critical long-term quality performance plans.

- 6 Sigma performers by 1995; and
- cycle time approaching two times the theoretical by 1995.

Policy and strategy at Rank Xerox

The first winner of the European Quality Award in October 1992, Rank Xerox is one of Europe's leading high technology companies providing a wide range of products, systems and services to handle customers' document and management requirements.

Similar to all previous cases, Rank Xerox, as noted before, believes that quality is the basis for driving the company forward. Quality principles are used for the development of policy strategy and all objective setting, from initiation to implementation and continuous improvement. The policy and strategy framework used by Rank Xerox relies on input from a wide variety of sources.

- Input from cusomers – this is their most important source of information. In addition to direct management contact, Rank Xerox measures customer satisfaction through local market research programmes implemented by individual operating units, and also conducts a range of Pan-European research programmes.
- Input from employees – all employees working directly with customers represent the second most important source of information. Through quality teams, a broad range of employees can contribute to the strategic development, implementation and refinement.
- Input from benchmarking – this is another source of information. Rank Xerox use benchmarking to deploy its leadership through quality strategy and they have extended it far beyond competitive analysis to include, for example, the identification and study of best-in-class organization for all areas of process management which are relevant to its business. Benchmarking is primarily used to identify what Rank Xerox needs to do in order to become the benchmark, and therefore the goals that it needs to set for itself.
- Input from suppliers – Rank Xerox has partnerships with hundreds of major suppliers who provide an essential source of information for establishing customer requirements and market developments. Their policy advocates open sharing of information with key suppliers, and so there is continuous promotion of sharing relevant business information and experience. In return, various suppliers give Rank Xerox feedback on their performance as a customer.
- External and societal trends – through studying community trends and benchmarking (for example, European pricing) and comparing distribution processes (with other competitors such as DEC, Olivetti and Apple), Rank Xerox has developed a strategy called 'the European dimension'. This clearly spells out the Rank Xerox approach to customer support, R&D, sourcing, marketing and distribution. In addition, Rank Xerox has developed a comprehensive environmental strategy.

- Business results – these are also a continuous source of information through regular review, and diagnosis of results contributes to strategic development and review. The basis for business planning is a process of policy deployment developed in 1989. Policy deployment, the annual process for turning strategic direction into operational business plans, is designed with quality principles in mind and allows Rank to:
  - establish customer requirements;
  - encourage people involvement;
  - involve people at all levels in the development and acceptance of objectives; and
  - establish management control mechanisms based on local empowerment within all the operating units throughout Europe, and voluntary regular self-assessment of performance, the outcome of which are used as an input to review the plans.

  Policy deployment has a great impact on strategic planning at Rank Xerox because it is the major vehicle for communicating the company's policy and strategy as well as giving feedback for further refinement and further developments. Review takes place on a regular basis; diagnosis of the results is based mainly on the use of customer satisfaction surveys which give an indication about the effectiveness of the strategy and the process modification which needs to take place in order to improve customer satisfaction. By being close to customers, their requirements for the future are clearly established and using input from benchmarking and competitive results are further inputs for the development of new strategies for the future. Through its business excellence certification programme, Rank Xerox encourages the use of self-assessment to establish progress and develop action plans at levels within the organization. Business excellence certification is used as a process to provide Rank Xerox with a continuous customer-based feedback on everything that is does. It is a vehicle that will ensure that Rank Xerox continues to improve standards of service, and to develop stretch goals for itself to remain a world leader in providing customer excellence.

It is very clear from the various examples covered that benchmarking is the key driver for strategic planning. It helps organizations establish the following:

- **Voice of the customer**
  By providing accurate information on customer needs and their levels of satisfaction with existing products and services provided, this input enables senior managers to develop strategies which will remedy situations where there are unfulfilled customer demands or where the levels of satisfaction are not acceptable. In addition, intelligence

obtained can enable senior managers to put together strategies that will cater for the future needs of its customer base.

- **Voice of the process**
  Benchmarking practices, technologies and methods, enable organizations to exploit state of the art knowledge and state of the art technologies, to establish superior ability and capability, to provide excellent services based on quality, cost, timeliness, delivery reliability, and responsiveness, amongst others.

5.4 The development of an effective benchmarking strategy

5.4.1 The starting point: having a clear vision

Any organization should be conscious of its mission, what it seeks to achieve and what mechanisms it uses to make sure its plans are on course. The development of a benchmarking strategy, therefore, has to start with:

- the purpose of the organization – the reason it is in the marketplace;
- a set of values – what the organization believes in and its guiding principles;
- a mission/vision – defining the role of the organization and what it seeks to achieve;
- objectives;
- strategies – for achieving critical success factors (CSFs);
- measures of progress on achieving CSFs; and
- action plans leading to desired business results.

5.4.2 A clear understanding of the business process

The conversion of mission/CSFs into realizable objectives demands a good understanding of the business process. Aspects of the business process which are key in leading to the achievement of CSFs are often referred to as critical processes (CPs).

Critical processes have to be managed properly and closely linked to CSFs so that business results can be achieved.

5.4.3 Using TQM as a competitive weapon

Benchmarking is the best tool for introducing improvements into the strategic planning aspects of any organization and for using TQM as a competitive weapon.

Benchmarking brings about the discipline of ensuring that:

- the strategic goals are attainable; and
- there is a strong link between the ends (CSFs) and the means (CPs).

Benchmarking takes the market/customer as a starting point and works inwards ensuring that the business process is capable of delivering consistently what the market/customer requires.

5.4.4 Linking benchmarking to the continuous improvement ethos

There are three major determinants of competitiveness:

- quality – product features, service features, customer care features;
- delivery – timeliness, reliability, consistency; and
- cost – total cost of delivery to the customer and incremental improvements.

Those competitive criteria can only be achieved by:

1. adopting a TQM approach for competing (as a means to drive the business process);
2. instigating a continuous improvement ethos; and
3. using benchmarking as a vehicle to direct the organization towards world-class competitiveness.

5.5 Strategic supremacy through benchmarking

Through the effective use of benchmarking in strategic planning, organizations can achieve short- and long-term targets of competitiveness. Using continuous improvement as the philosophy and a plan–do–check–act (PDCA) wheel of improvement, strategic planning can be reviewed on a regular basis, its elements monitored and changed to ensure effectiveness is achieved all the time. The use of benchmarking can be deployed at all levels within the organization and this will produce a dynamic process of managing at all levels.

At senior management level the concern will be with strategic issues. Through the use of PDCA, business goals and objectives can be developed from a thorough understanding of customer requirements, market conditions, and organizational capability. These plans can then be communicated to all levels, for performance to take place.

At an operational level, benchmarking can ensure that core activities or processes are managed to produce quality and benefits for the end customer. Once again, through the use of PDCA processes, quality and benefits can be optimized for the end customer.

Strategic planning and the implementation process have to be an integral exercise. Benchmarking can provide the required discipline for aligning people and ensuring there is a common focus on the end customer.

- Benchmarking ensures that the key elements used for strategic planning are all used for the benefit of the end customer. It enables senior managers to deploy resources in the right way and to prioritize goals and targets. Through accurate information on customer requirements and market conditions, resources can be deployed in the right direction. Benchmarking gives clear signals on threats, but also on opportunities, and as a result of the intelligence obtained, senior managers can react correspondingly.

There is therefore a distinction between strategic development for a defensive reason, and strategic development for exploiting opportunities and influencing market conditions. In relation to this point, Ohmae (1982) argues:

> when one is striving to achieve or maintain a position of relative superiority over a dangerous competitor, the mind functions very differently from the way it does when the object is to make internal improvements with reference to some absolute model. It is the difference between going into battle and going on a diet.

- Having identified where the critical success factors are, organizations have to develop means to achieve them. Therefore they must understand where the critical processes are and how they are managed and controlled. Determining organizational capability becomes fundamentally important.

  To minimize the risk in decision making, and to develop effective strategies, the voice of the customer and the voice of the process must be listened to. Capability determines superiority in the marketplace and can establish true differentiation. As argued by Ohmae (1982):

  > If you are fighting with a competitor who has equal qualifications, effective and persistent execution in critical functional areas may be the only differentiating factor. Toyota's persistence in routing out waste from its organization and Hitachi's corporate wide management improvement (MI) activities are good examples of doing much better in areas everyone deals with anyway.

- Senior managers rely on information to make effective decisions. It is their task to harness creativity and optimize synergy in order to achieve the desired levels of competitiveness.

  Through the use of benchmarking, complacency can be avoided and continuous improvement becomes the norm. It is only by consistently asking why, that organizations can move from closing negative gaps to establishing strengths, consistency and superiority in all aspects of their delivery to the end customer. They will gradually move from a mode of continuous improvement to being a continuous learning organization. In relation to this point Ohmae (1982) argues:

  > The strategists method is very simple. To challenge the prevailing assumptions with a single question: Why? and to put the same question relentlessly to those responsible for the current way of doing

things until they are sick of it. This way bottlenecks to fundamental improvement are identified, and major breakthroughs in achieving the objectives of the business become possible.

- Successful organizations use quality management principles in a pro-active manner. They have learnt to move away very quickly from a mode of fire-fighting, dealing with negative issues, and being absorbed internally with closing negative gaps, to become organizations that seek to establish superiority by developing full packages of additional features that will delight the customer.

  Benchmarking indicates the right way for organizations to go beyond providing basic requirements for the customer. Strategic planning should not be concerned with the development of defensive policies, but should be used to establish superior performance standards. Benchmarking provides intelligence, information, data and knowledge for senior managers to encourage them develop plans that will lead to better performance standards. As Ohmae (1982) explains:

  > Strategic thinking in business must break out of the limited scope of vision that entraps deer on the highway; it must be backed by the daily use of imagination and by constant training in logical thought processes. Success must be summoned: it will come unbidden and unplanned. Top management and its corporate planners cannot sensibly base their day-to-day work on blind optimism and apply strategic thinking only when confronted by unexpected obstacles. They must develop the habit of thinking strategically, and they must do it as a matter of course. Ideally, they should approach it with real enthusiasm as a stimulating mental exercise.

- Benchmarking produces the discipline for translating strategies into tangible outputs for the benefit of the customer, by referring the goals to the processes and the process capability. It places great emphasis on what to do, how to do it, when to do it and more importantly on performance measurement and the output. As Ohmae (1982) argues:

  > Corporate performance is the result of combining planning and execution. It resembles a boat race. No matter how hard each crew member rows, if the coxswain doesn't choose the right direction, the crew can never hope to win. Conversely, even if the coxswain is a perfect navigator, you cannot win the race unless the rowers strive hard in unison.

  Benchmarking is essential for the conversion of strategies into realizable goals.

6

Types of benchmarking

Benchmarking began as an in-depth, ongoing study of best competitors, a detailed reverse engineering of competitor products, technology processes, what they achieved and how they did it. Operating capabilities and features of competing products were subject to a tear down analysis. Most, if not all, of this early work, in the late 70s and early 80s, was conducted by Xerox and Rank Xerox simply to establish what needed to be done in order to stay in business. As has been described earlier, data gathered about direct competition revealed just how bad the situation was. It is well documented that an increasingly tight focus on all aspects of direct competitor manufacturing methods, and so on, enabled the company to slow down and eventually turn the tide.

Bearing in mind, then, from today's perspective, as benchmarking is all about searching out the best of the best in order to bring about superior performance, there are some things worthy of note about benchmarking against direct competitors, i.e. competitive benchmarking.

6.1 Competitive benchmarking

Competitive benchmarking, as its name implies, can be used as a way of informing people how badly or well they are doing against direct competition. The main disadvantages are that it can be difficult, even impossible, to obtain information on competitive processes or targets. Customer feedback obtained through anonymous surveys conducted by independent consultants can provide excellent objective indicators of company versus competitive performance and is one method used to compensate for this disadvantage. However, whilst it is helpful to compare against similar industries, a 'me too' strategy may be the response. Breakthrough actions might not be achieved. Care is also required to ensure that competitors are truly comparable.

An example of competitive benchmarking would be Xerox benchmarking its low volume printers time to market against that of Canon, Minolta, Sharp, etc. with the clear intent of understanding who is fastest and more importantly, how do they do it? The very heart of benchmarking is understanding how better results are achieved and adapting the 'how' into one's own organization so as to overtake the competition in terms of results.

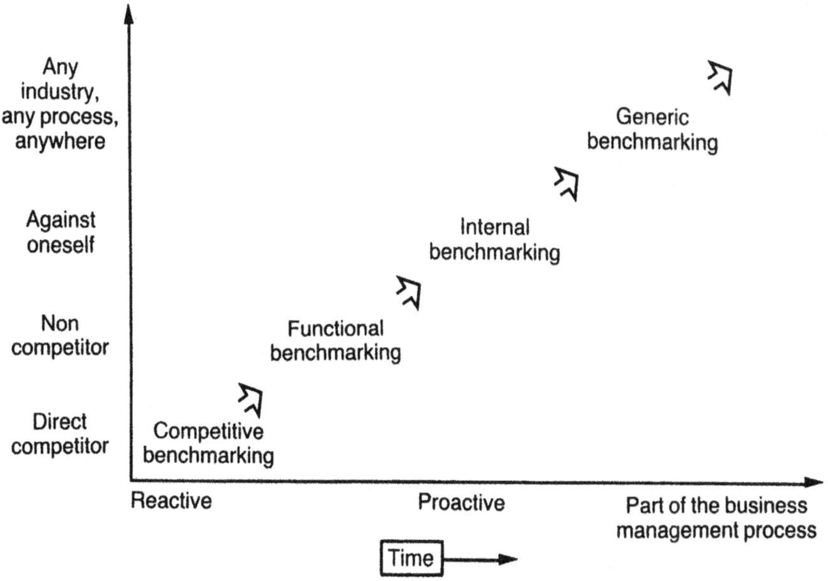

Figure 6.1 The evolution of types of benchmarking.

As the understanding of this notion of externally focused improvement began to be better understood, the realization began to dawn that the technique could possibly be used in a non-manufacturing environment and the phrase 'Functional Benchmarking' was coined.

6.2 Functional benchmarking

Functional benchmarking compares specific functions e.g. distribution, logistics, service, etc. with best-in-industry and best-in-class. One major advantage of this approach is that it is easier to gain access into non-competitive organizations as it is less threatening and there is also a greater likelihood that a two-way partnership can be forged with a greater potential for learning.

It is encouraging how open a business will be if asked to share success stories; this is particularly true when the benchmarking focus is on processes. Most organizations seem willing to share information if approached in a professional manner and if they know the purpose of the questions.

In addition, a source of inspiration from another company can challenge existing assumptions and lead to new approaches. There is however the limitation that, because it only relates to specific functions, it may not be of benefit to other operations in the business concerned.

Additionally, care is required in selecting companies to benchmark as the nature of comparison is complex. Cultural or demographic factors, for example, and differing definitions of measurement frequently play a key role in undermining the credibility and subsequent implementation of the findings of such studies.

The following is an example of functional benchmarking from the mid-80s. Rank Xerox Limited in Europe had identified a need to improve radically a specific part of its distribution and logistics function. It identified and subsequently worked with, benchmarking partner companies as diverse as Volvo, 3M, Ford, IBM and Sainsbury's. The inclusion of a supermarket chain as a benchmarking partner to an office business systems company was quite radical as the notion of benchmarking detailed processes or sub-processes was little understood at the time. There was some marked reluctance from people who could not understand what photocopiers had to do with cabbages! However, these difficulties were overcome and the outcome of this one exercise enabled a significant reduction in inventory levels without reducing the level of service to customers.

Another well known example of functional benchmarking which is dealt with in great detail in Bob Camp's book is the story of Xerox Corporation in the US benchmarking part of its distribution operation against that of a mail-order catalogue company called L. L. Bean, again with significant improvements being obtained.

6.3 Internal benchmarking

Benchmarking does not have to mean comparison with another company. For many multinationals, an intensive internal search is the starting point for any benchmarking exercise. It is the continuous effort of establishing good practice in the various operations of the overall business.

For example, Rank Xerox can usefully compare manufacturing at Mitcheldean (UK) against Rank Xerox manufacturing at Venray (Holland) and Lille (France). This could be any part of the operation, manufacturing or non-manufacturing.

For large organizations which are spread across many countries there has to be an opportunity to compare or to establish where the benchmark is by function or by business process.

Examples might be, say, the way in which service engineers respond to customer requests for repairs. Which part of the organization responds best from the viewpoint of the customer? Which part of the organization handles complaints most efficiently? What do they do that makes them the best – the benchmark?

Internal benchmarking, although it sounds relatively easy, can be extremely difficult to apply. How can an objective comparison be made where on the face of it, cultural differences might preclude it if, for example, a company operates in Finland, Greece and Germany?

Another point to bear in mind about internal benchmarking is that even if the potential barriers already identified can be overcome, there is a risk that any targets set as a result of an internal benchmarking study could be set against an internal standard, which may or may not be the world class standard which we are seeking.

There are ways of overcoming 'cultural' barriers. The Malcolm Baldrige model in the US gives a few pointers, as does the European Foundation for Quality Management (EFQM) model. Rank Xerox has developed a model called Business Excellence Certification (BEC) based on the Baldrige/EFQM models and has used it successfully to demonstrate that comparisons between countries are in fact very useful when approached in a consistent, well-documented way. (Chapter 9 deals with this in more detail.)

6.4 Generic benchmarking

As the understanding and increased use of benchmarking in the different forms previously described grew, it gradually became apparent that the wider, more generic the approach to the whole subject of benchmarking then the more innovative could be the results.

Generic benchmarking is similar in many respects to functional benchmarking except that it focuses on multifunctional business processes – the very processes that are at the heart of the business. Once the critical business processes have been identified these can become the subject of benchmarking against any organization regardless of size, industry, or marketplace, providing that similar, generic processes exist there. It is undoubtably the latest evolutionary stage of the art of benchmarking and can apply to any area of a business operation. It is the state of mind of an organization which encourages the continuous effort of comparing functions and processes with those of best in class, wherever they are to be found.

However, until an organization really understands and has gone through the evolutionary process of understanding and gaining experience and business benefit from competitive, functional and internal benchmarking then it is unlikely that the mind-set of the organization is capable of maximizing the potential of generic benchmarking.

One very recent example was a large company making large, technically complex products benchmarking its 'marketing to manufacture' process against a company with a similar process but in the fast-moving consumer goods marketplace.

The most important point to remember about generic, and indeed any of the other types, is that the process, not what goes through the process, is critical when thinking about possible benchmarking partners. Once this concept is understood and accepted in the organization then the sky really is the limit in terms of the benefits that benchmarking can bring.

7

Benchmarking, benchmarking processes

7.1 Introduction

Many managers strongly believe that successful benchmarking exercises are only the result of choosing carefully the right methodology. The process of benchmarking, however, although being a tool, is very difficult to apply. It challenges the existing culture of work and scientific practices and methodologies in place. In addition, successful outcomes from benchmarking are heavily dependent on the presence/absence of a commitment to continuous improvement and a TQ-based culture.

The Rank Xerox benchmarking methodology (comprising five stages of planning, analysing, interrogation, action and scrutiny, and ten steps) is the best known and perhaps most widely used one. In most recent years, other methodologies have been suggested; some are based on incremental modifications of the Xerox approach, while others claim to be different.

This chapter compares and contrasts the best known benchmarking approaches based on a list of key criteria which are thought to be essential.

7.2 Criteria used for benchmarking

Seven criteria were identified for this benchmarking exercise.

1. *Strategic focus* Benchmarking processes have to lead to the setting of objectives and stretch goals based on a thorough understanding of customer requirements and process capability.
2. *Operational focus* Whilst benchmarking processes are about determining gaps in performance and setting clear objectives, it is absolutely essential that they are translated into practice, through a focus on the process and its performance, at an operational level.
3. *Customer focus* Market-driven strategies are those strategies which focus entirely upon the customer. For a benchmarking process to be successful, it has to set clear customer-based targets and help drive the performance of processes for optimizing levels of output which will represent value for the end customer.
4. *Process focus* Benchmarking is only meaningful as a process if it focuses on the process or the activity task. Many benchmarking

exercises focus too much on the outcomes (i.e. the metrics, the absolutes) without a clear understanding of what takes place, how it happens, why it is done in a certain way, etc. Processes reflect the strength of an organization (i.e. its capability to deliver, its flexibility to respond, etc). Such knowledge is crucial for setting stretch targets.
5. *Link to TQM* Benchmarking is an integral element of the total quality management philosophy. It has a similar approach to internal methods of problem solving and quality improvement.
6. *Continuous improvement (PDCA)* Since benchmarking is an integral element of TQM, it has to reflect a culture of continuous improvement. Process-based benchmarking exercises have to be repeated on a regular basis to strengthen the process further.
7. *Continuous learning* Benchmarking is about newness and innovation. It is doing the right things (determined by the clear establishment of market needs) and doing them right (by inspiration from leading organizations and through a control of capability of procedures). Benchmarking is a futuristic approach; it is concerned with the next set of objectives.

7.3 Methodolgy of benchmarking

Fourteen benchmarking procedures were considered in this comparative exercise. These methodologies are based on actual usage in companies and also prescriptive approaches. The comparisons were based on the seven key criteria described above, and a rating scale of 1–3 was used to allocate scores for each criteria.

- A scale of 3 reflected a strong link between methodology and key criteria. The link must be explicit, either as an actual step in the process or a prerequisite for effective benchmarking.
- A scale of 2 reflected a moderate link. The link could be implicit (i.e. use of quality improvement tools imply a link with TQM).
- A scale of 1 reflected a weak link; if no link is inferred, it will be assumed that the factor concerned is regarded as non-essential.

This benchmarking exercise was conducted using the limited amount of information available; in some instances, where no score was given, this is not necessarily intended to mean 'no link' but rather 'no information'.

7.4 The benchmarking methodologies considered

Xerox (Robert Camp)

This entails five phases involving ten steps.
Planning
1. Identify what is to be benchmarked.

2. Identify comparative companies.
3. Determine data collection method.

Analysis

4. Determine current performance 'gap'.
5. Project future performance levels.

Integration

6. Communicate benchmarking findings and gain acceptance.
7. Establish functional goals.

Action

8. Develop action plans.
9. Implement specific actions and monitor progress.
10. Recalibrate benchmarks.

Maturity

- Leadership position attained.
- Practices fully integrated into processes.

Comments

This is perhaps the best known and documented methodology as befits the company that undertook much of the pioneering work in benchmarking.

Strategic/operational focus Both are emphasized. The methodology requires that benchmarking be integrated into the corporate planning process. Processes, products, services, etc. can all be benchmarked but prioritization is important. Typically, this relies upon cascading the mission statement down through strategies and customer–supplier chains to lower level processes and deliverables. These become candidates for benchmarking. Little distinction is drawn between strategic and operational focus.

Customer focus Benchmarking must be externally focused or target setting and planning are forced to rely upon unsatisfactory extrapolation of historical performance and targets. In real terms, this means the focus is upon the marketplace and the customer.

Process focus Not exclusively! Benchmarking can also be used for products, services, etc. However, process-based benchmarking predominates. This recognizes that 'customer satisfaction' is the output from the overall business process itself, built up of many other processes. All other things are themselves process outputs.

Link to TQM There is no explicit link made. However a key success requirement is the involvement of process owners in benchmarking. This requires an organization ready, willing and able to accept change (without fear). It is argued that this is more likely with an effective TQM programme. The use of basic quality tools is advocated in the methodology.

Continuous improvement (PDCA) Benchmarking is a continuous process, not a one-off exercise. The methodology is externally focused –

the external environment changes. The facility to review both benchmarks and the benchmarking process is provided in the methodology.

Continuous learning This is the ultimate output of the benchmarking process, as represented by the maturity phase of this methodology. In such an organization, benchmarking is fully integrated with the planning process and is 'business as usual'. It is performed at all levels, by all, continuously; generally initiated by affected individuals (process owners), it may be regarded as institutionalized.

Post Office Counters Ltd (POCL)

The Post Office adopts a practical approach to benchmarking. Their methodology is explicit not only about what is meant to be done, but on how it is to be done.

Strategic/operational focus Their methodology does not require that benchmarking be integrated into the strategic planning process. Focus is therefore not explicitly strategic; it is predominantly operational.

Customer focus The only 'rational' criteria for process selection given is 'customer feedback on current processes'. From this standpoint, then, there is some customer focus. However, as focus is generally missing from this methodology, one could reasonably expect explanation of how customer feedback and process selection are actually related.

Process focus This methodology is very strongly process-based. It is recognized and accepted that processes are at the heart of benchmarking.

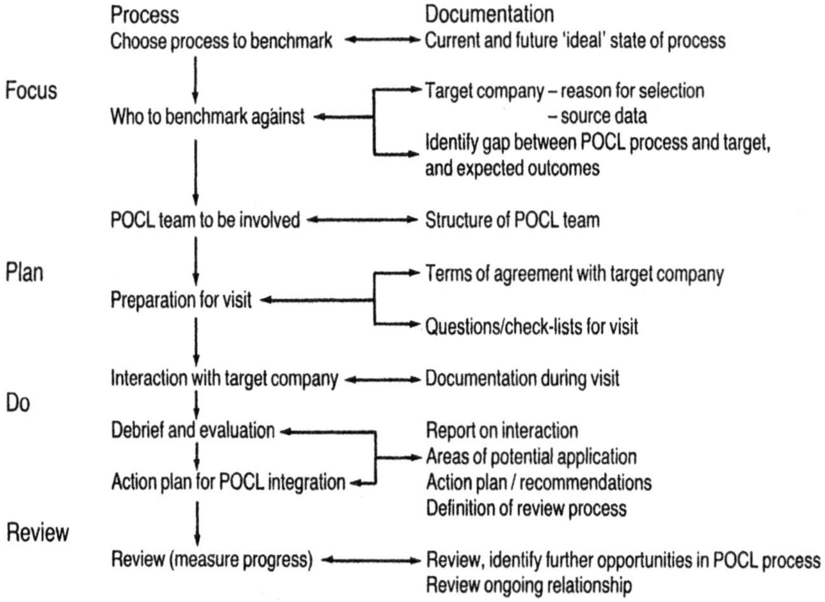

Figure 7.1 Post Office Counters (POCL) methodology

Without processes, it is not possible to transcend industry boundaries (i.e. beyond the limits of products and other specific process outputs).

Continuous improvement (PDCA) This methodology contains a review phase, and to some extent is constructed around the improvement cycle. However, details of how the review is structured are not given.

Royal Mail

The Royal Mail and Post Office Counters Ltd share the same benchmarking methodology. Additionally, there is a link with TQM when the methodology is employed by the Royal Mail. Benchmarking was developed within this company using a pilot study based upon the TQM process. It is thus reasonable to assume linkage and interdependence of benchmarking and TQM in this case.

International Benchmarking Clearing House (IBC)

The following methodology contains four basic phases and up to 36 steps. It is not prescriptive; these steps must be regarded as possible steps. This methodology must be adapted, intelligently, to the needs of the organization.

Planning a study

- Select process
- Gain process owner's participation
- Select leader and team
- Identify customer expectations
- Analyse process flow and measures
- Define process inputs and outputs
- Document process
- Select CSFs to benchmark
- Determine data collection
- Develop a preliminary questionnaire

Collecting data

- Collect internal data
- Perform secondary search
- Identify benchmarking partners
- Develop a survey guide
- Solicit participation of partners
- Collect preliminary data
- Conduct visit

Analysing data

- Aggregate data
- Normalize performance
- Compare current performance to data
- Identify gaps and root causes
- Project performance to planning horizon
- Develop case studies of best practice
- Isolate process enablers
- Assess adaptability of process enablers

Adapting and improving

- Set goals to close, meet and exceed gap
- Modify enablers for implementation
- Gain support for change
- Develop action plans
- Communicate plan
- Commit resources
- Implement plan
- Monitor and report progress
- Identify opportunities for benchmarking
- Recalibrate the benchmark

Benchmarking according to this methodology is a cyclical process which emphasizes a number of issues.

- Benchmarking is not the same as competitive analysis. Benchmarking, being process-focused, is able to transcend industry boundaries.
- A culture open to change is important to benchmarking success, it cannot flourish in the absence of TQM.
- Management support and integration with the strategic planning process are fundamental.
- Planning, organization and understanding one's own process are crucial.

Strategic/operational focus In selecting candidates for benchmarking one must answer the question 'what does it take to win in this business?' However, benchmarking is appropriate to any and all processes. It can be both strategically and operationally focused, this being a matter of prioritization. Priorities are decided by considering customer needs, strategic needs and their overlap with processes.

Customer focus, being process-based, linked to TQM and the improvement cycle are all considered as necessities for effective benchmarking.

Vaziri

H. K. Vaziri's methodology emphasizes the need to make a clear decision on what to benchmark. This can be achieved using quality function deployment or similar techniques.

Strategic/operational focus Vaziri's methodology (Fig. 7.2) is essentially strategically focused. It recommends use of QFD, etc. to consider organization mission, strategic direction and customer needs to set benchmarking priorities. The aim being to identify the 20% of 'customer outputs' that account for 80% of the desired business improvement. The processes that yield these outputs are the ones to benchmark.

Vaziri recognizes that benchmarking is a component of an overall TQM programme. Through customer focus, ideas can be generated from any best practice source, thus allowing an organization to meet and exceed customer needs. Benchmarking is seen as the way to manage process innovation.

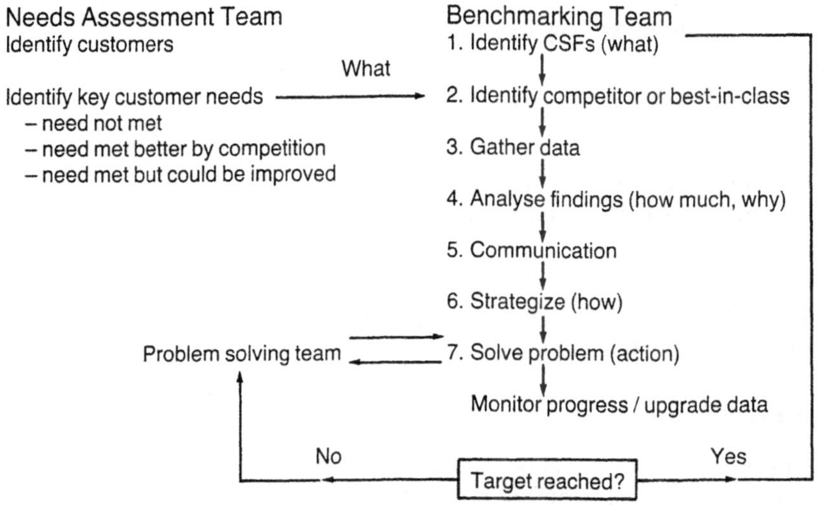

Figure 7.2 **Vaziri's methodology**

Price Waterhouse

Identify

1. Identify functions to be benchmarked.
2. Identify key performance variables to measure.
3. Determine whom to compare against.
4. Establish data collection methodology.

Analyse

5. Measure our own performance.
6. Measure performance of competitors and best practice leaders, determine gaps and reasoning.

Plan

7. Communicate findings and gain acceptance.
8. Develop action plans.

Implement and evaluate

9. Implement actions and monitor progress.
10. Recalibrate benchmarks over time.

Comments

The Price Waterhouse methodology is very similar to the Xerox benchmarking work (amongst others). There has been little novel contribution from Price Waterhouse itself. Having said this, what is stated is valid and some valuable points are reiterated.

Strategic/operational focus Benchmarking can be used at both operational and strategic levels. The decision on what to benchmark must be made with reference to what impact is made upon customer satisfaction and what functions are key to business strategy.

Unfortunately it is implied that there is a choice between business function or process. This fails to note that:

- many processes cut across functional boundaries – therefore functional focus might result in artificial interfunctional barriers; and
- process focus is essential if comparisons are to be easily made beyond one's own industry.

Link to TQM This is not explicitly covered; however it is recognized that skills, ability and willingness of an organization to change is of importance to effective benchmarking.

McKinsey & Co.

1. Choose what process to benchmark.
2. Select key measures and practices (both input and output measures on quality, timeliness and cost axes).
3. Identify comparable processes and benchmark companies.
4. Assess the world-class approach.
5. Develop change priorities.

Strategic focus The McKinsey approach is strategically focused. It is recommended that the benchmarking organization take a strategic overview of itself; in doing so, it should consider which areas of the business (and thus processes) contribute to the greatest proportion of 'added value'. This is not always obvious; for example, a manufacturing firm using many bought in assemblies might find its supply chain management process covers most of the value added.

Customer focus This is not explicitly covered. However, if 'value added' refers to 'value to the customer' then customer focus is implicit in process selection.

Process focus This methodology is strongly process-based. It states this as the key superiority of benchmarking over competitive analysis.

Codling

In the methodology devised by S. Codling, benchmarking is comprehensively detailed. It covers the essential aspects of theory but is very practical in nature.

Strategic/operational focus The methodology is essentially strategic in focus. However, it ultimately must be devolved to the operational level. It requires that a 'subject area' is chosen by considering an organization's mission statement, corporate goals and core activities. Whilst this is done the question asked is 'what must be improved to really impact corporate goals?' – essentially what are the organization's critical success factors?

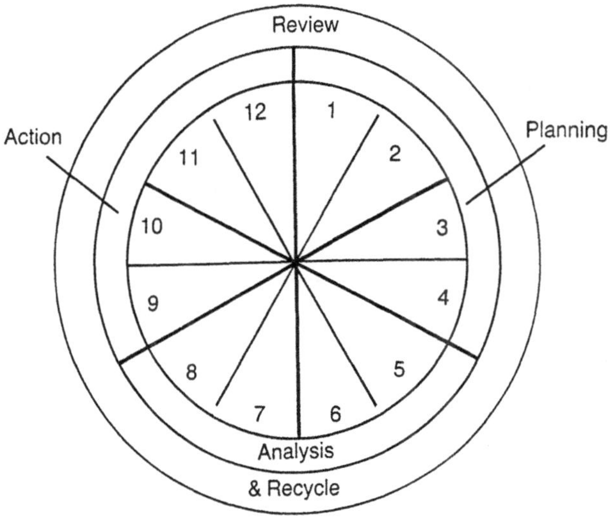

1. Select subject area; 2. Define the process; 3. Identify potential partners; 4. Identify data sources; 5. Collect data: select partners; 6. Determine the gap; 7. Establish process differences; 8. Target future performance; 9. Communicate; 10. Adjust goal; 11. Implement; 12. Review: calibrate.

Figure 7.3 Codling's methodology

Once chosen, the area is broken down in a similar fashion to the process/operational level and thus critical processes are identified for benchmarking.

Customer focus Benchmarking is viewed as being a tool to take organizations from being 'production-led' to being 'customer-driven'. Customer focus is therefore viewed as an output of benchmarking rather than a prerequisite.

Process focus Unlike traditional competitive analysis, benchmarking is not concerned with outputs (products, etc.) but with processes. Processes deliver outputs, are reproducible beyond industry boundaries and thus facilitate development of best in class practice.

Link to TQM/PDCA Codling argues that benchmarking can be performed whether or not an organization is going down the TQM route. However, in practice, the more quality is ingrained within the organization the more readily people will relate to benchmarking. This methodology relates itself directly to TQM and the continuous improvement cycle.

Learning organization Benchmarking is not a one-off event; it is continuous and strategic in nature. The ultimate aim is an organization within which benchmarking is just another facet of the culture, conducted by all at all levels.

McNair & Leibfried

Strategic/operational focus In the C. J. McNair and K. H. Leibfried methodology, as with a number of other approaches, a strategic focus is used to orientate the benchmarking process. Action, however, can only occur when one considers the more detailed or operational aspects.

Identify core issues → Internal baseline data collection → External data collection → Analysis → Change Implement →

Input

Issue – Unmet customer needs – Performance gap – Problem areas – Strategic advantage	Overview of process Current measures Potential drivers and external organizations	Benchmark questionnaire	Compare and contrast benchmark data	Implementation plan Issues

Output

Defined benchmark area Overview of key processes to benchmark Selected performance measurements Identify potential drivers and external organizations	Process flow mapping Validate drivers Benchmark target companies Short-term operational improvements Benchmark questionnaire	External company(s) Process analysis performance assessment and measures	Gap Process improvements/ reengineering opportunities New – Flows/policies/ procedures Implementation plan Outstanding issues	Plan to close gap Action to close gap Recalibrate benchmarks Additional analysis/ benchmarking to address issues

Figure 7.4 McNair and Leibfried methodology

In recognition of the limited nature of benchmarking resources it is proposed that a rational selection method is used for benchmarking targets. One must consider the organization's CSFs (derived from the usual mission-goals-strategy deployment). In doing this one asks 'what is it that will most impact stakeholder value?' The answers become benchmarking candidates.

Customer focus Benchmarking is given direction by an organization's goals and objectives. Implicit in these should be an understanding of customer needs. Therefore, the customer focus of benchmarking is largely determined by an organization's strategic direction. More explicitly, it is recommended that external 'best practice' should be evaluated in terms of how this is seen from the customer's perspective.

Process focus It is possible to focus benchmarking on roles, processes or strategic issues. Thus benchmarking need not be entirely process-based. The purpose of benchmarking is to evaluate the way:

- a role is performed (what);
- a process is undertaken (how); or
- a strategic issue is defined (why).

Link to TQM The methodology sees benchmarking as an ideal complement to TQM. However, TQM is not regarded as an explicit prerequisite to benchmarking. TQM pursues continuous improvement by discovering the 'root causes' of problems – the whats. Benchmarking complements this by outlining possible directions for change – the how. Thus benchmarking is able to transform 'customer-orientated' to 'customer-driven'.

AT&T

First things first

1. Determine who clients are (process owners and planners).
2. Advance the clients from literacy to champion stage.
3. Test the environment. Identify commitment, expose barriers.
4. Determine urgency. Avoid states of panic or apathy.
5. Determine scope and type of benchmarking required.
6. Select and prepare team.

Process

7. Overlay benchmarking process onto business planning process.
8. Develop benchmarking plan.
9. Analyse data.
10. Integrate the recommended actions.
11. Take action.
12. Continue improvement.

Strategic/operational focus The methodology was developed when AT&T were faced with the strategic dilemma of market deregulation. Consequently, the methodology is intended to be strategically focused. Process selection is based upon an understanding of mission, strategic direction and how this cascades down to the process level.

Customer focus The need for an external focus is stressed but customer needs are never explicitly covered. We are left to deduce concentrating upon best practice guarantees that customer needs are met and exceeded.

Link to TQM There is a strong, although not explicit, link between this benchmarking methodology and TQM.

Learning organization The methodology relies heavily upon 'grafting' a specialist benchmarking team onto existing process owners (clients). Ideally, process owners should ultimately initiate benchmarking themselves.

Alcoa

1. Decide what to benchmark (process owner identifies and checks for relevance to customer needs, strategic considerations and business needs).
2. Planning the benchmarking project (team and proposal).
3. Understand own performance.
4. Study others (identify candidates, select, prepare for visit and conduct).
5. Learn from data (identify performance gaps and underlying practices).
6. Use the findings.

The methodology is strongly strategically focused, requiring consideration of mission and strategic direction prior to benchmarking candidate selection. A process basis is implied although not to the exclusion of other approaches. The methodology generally accepts that benchmarking cannot succeed in absence of TQM.

NCR

1. Identify key performance measures.
2. Identify best-in-class companies.
3. Measure performance of best-in-class companies.
4. Measure own performance and identify opportunity gaps.
5. Specify programmes and actions to meet/surpass the best-in-class.
6. Implement and monitor results.

TNT

Understand process
- Adopt the 'blank sheet of paper' approach.
- Analyse the process.
- Identify area of concern.

Establish present position
- Identify world-class players and areas of excellence.
- Carry out comparison.

Formulate and implement plans
- Establish strategy.
- Allocate resources.
- Formulate plan and timescales.

Continuous assessment and improvement
- Establish performance monitors.
- Review and reassess progress.
- Reassess (data and result, take action).

Review and reassess
- Schedule regular review meetings.

Schmidt

J. A. Schmidt's methodology involves the following.

1. Research own company first and identify key success factors of the business.
2. Establish scope and basis of benchmarking comparisons: strategic customer, cost versus best meets company's improvement needs.
3. Select the group of comparators; identify and screen companies for the comparison group.
4. Develop a detailed plan for data collection and processing: sources could include customers' trade associations, employees, joint venture partners, co-operative data exchanges, interviews or surveys.
5. Develop conclusions and establish performance targets to ensure that benchmarking is a catalyst for the future.

7.5 Outcome of benchmarking exercise

The comparison of the 14 various benchmarking processes considered suggests that there are many similarities between the different approaches used. The slight differences suggest perhaps:

- better clarity,
- explicit focus on each criteria,
- logical progression, and
- completeness.

The Rank Xerox and Codling approaches appear to be the most complete ones, although the latter appears to be heavily based on the former.

Table 7.1 illustrates the various benchmarking scores for each methodology, represented graphically in Figure 7.5 which illustrates the difference between the various methodologies under each individual criteria.

Table 7.1 Methodology scores

Methodology	Strategic Focus	Operational Focus	Customer Focus	Process Focus	Link to TQM	Continuous Improvement	Continuous Learning	Aggregate
Xerox	2	2	3	2	2	3	3	17
PO Counters Ltd	1	3	1	3		2		10
Royal Mail			1	3	2	2		8
IBC	2	2	3	3	3	3	2	18
Vaziri	3		3	3	3	2		14
Price Waterhouse	2	2	2		1	3		10
McKinsey & Co	3		1	3	1			8
Codling	3	2	2	3	2	3	2	17
McNair & Leibfried	3	2	2	2	2	3		14
AT&T	3		1	3	3	3		13
Alcoa	3		3	1	3			10
NCR								
TNT				3		2		5
Schmidt	3		3					6
Aggregate	28	13	25	29	22	26	7	

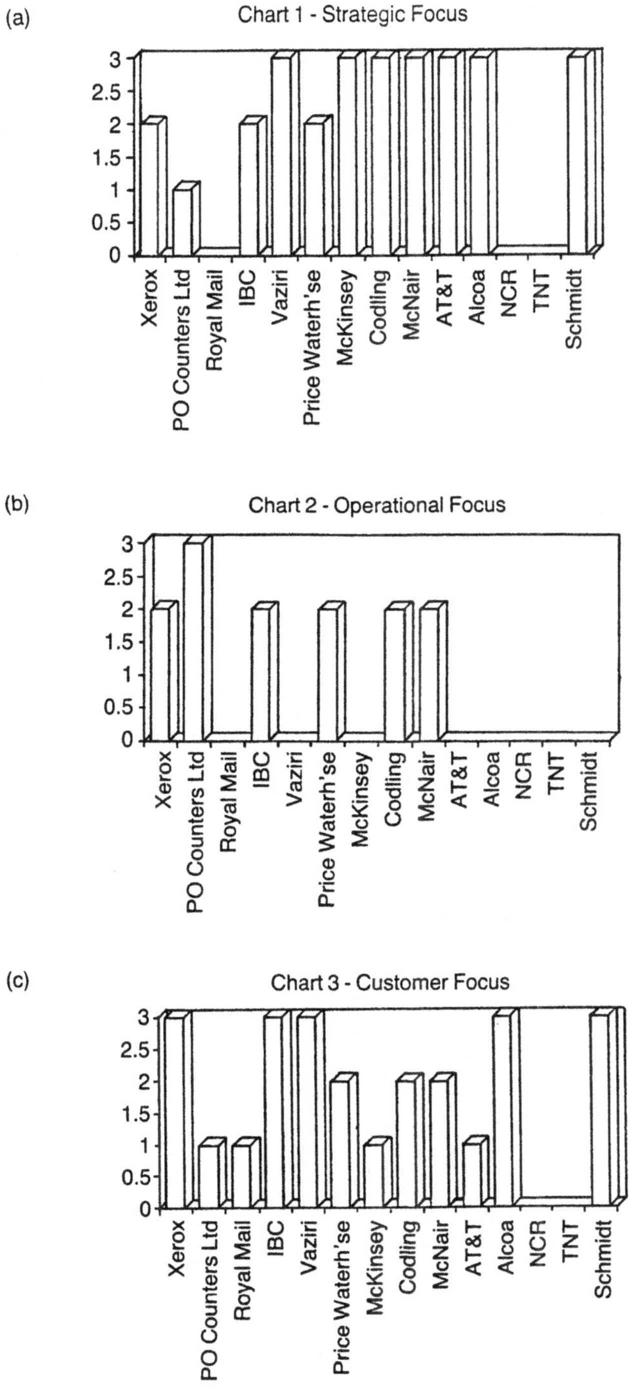

Figure 7.5 (a) Strategic focus; (b) operational focus; (c) customer focus.

(d)

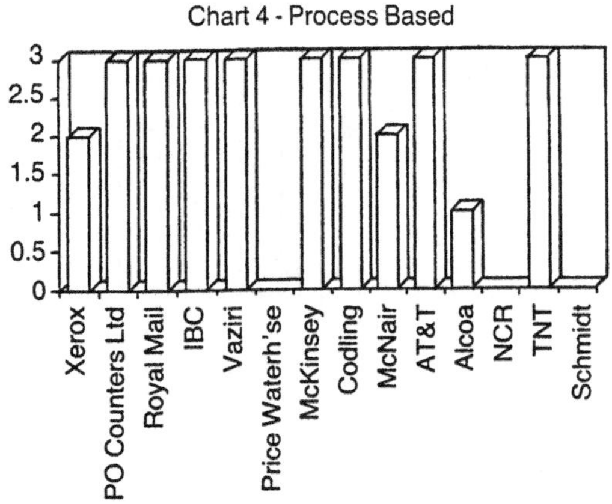

(e)

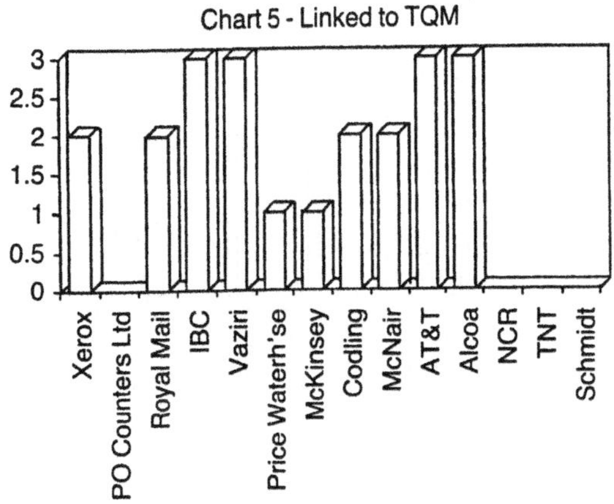

(f)

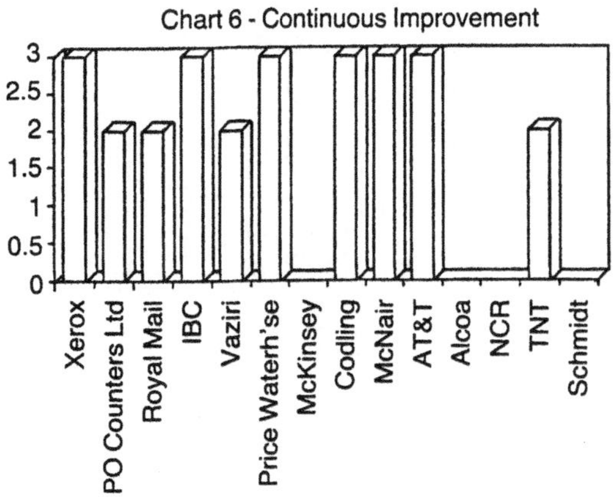

Figure 7.5 (d) Process based; (e) linked to TQM; (f) continuous improvement.

(g)

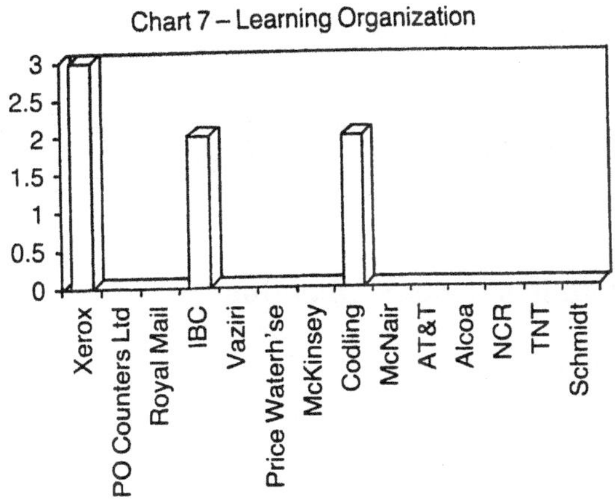

(h)

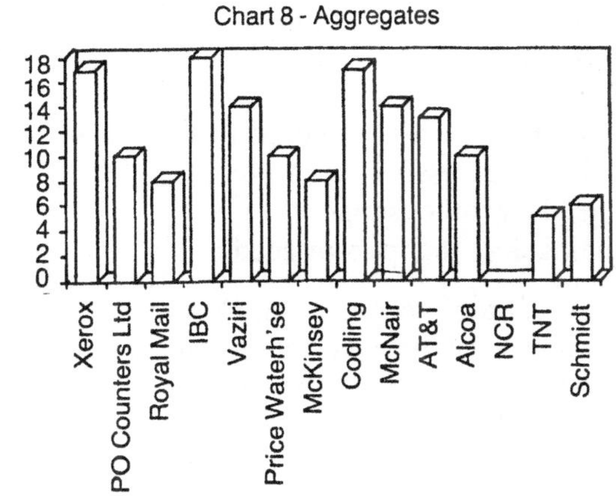

(i)

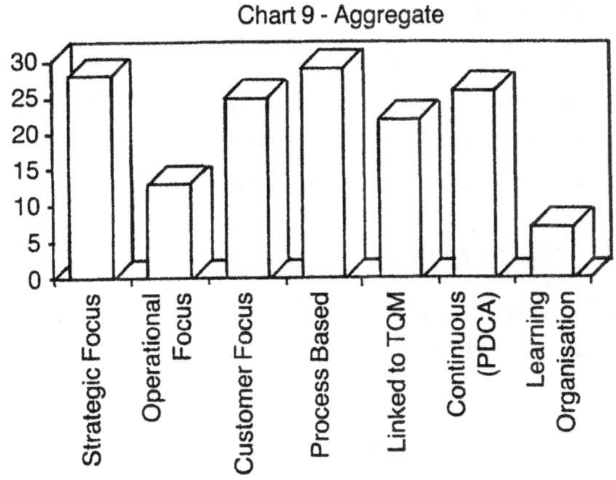

Figure 7.5 (g) Learning organization; (h) aggregates; (i) aggregate.

8

Data collection approaches

8.1 Prerequisites to data collection

8.1.1 Training

Once a benchmarking study team has been formed some questions need to be answered.

- What initial training on the subject of benchmarking is needed?
- What additional training is required?
- What start up training is needed in order to get the team up and running?
- What ongoing support is required to keep them running and to help facilitate their progress?
- Do they need any other support?
- What training does the sponsor need?

The sponsor certainly needs to know something about benchmarking. He needs to know what the expectations are, the length of time the project is likely to take, etc. He needs to know how to inspect progress. He needs to feel confident that the analysis of the benchmarking data collection exercise is accurate, if he is to be convinced that any proposed change is the right thing to do.

8.1.2 Communication

People who are going to be impacted by any proposed change to current activities need to be advised well in advance. They need to understand what the impact of any planned change will have on them – on a day-to-day basis, on their work profiles, and on their working life in general.

8.1.3 Skills requirements

During the course of the exercise, the team, as a whole, will need the following skills/knowledge:

- project management
- interpersonal skills
- consulting/influencing skills

- listening skills
- flowcharting techniques
- analytical skills
- presentation skills
- business writing
- time management/diary planning
- action planning
- questionnaire development
- interviewing skills
- telephone techniques
- facilitation skills
- problem solving
- selling skills
- verbal and written communication skills

Whilst it is not necessary for each team member to have a thorough knowledge of all of the above, it is very important to the success of the project that all the skills are present in the team.

8.1.4 Process mapping abilities

The team is going to have to work together, looking at a particular area of the business. They are going to put together a data collection methodology, and collect information from other organizations. They are then going to have to analyse that information, and present the findings of that analysis to people within their own organization, to internal customers, to managers, and to any other departments who have a need to know, especially if recommendations for change are going to be brought about.

The team has to be able to map the process, or sub process exactly as it currently exists with all the activities, decision points, inputs and outputs, so that a true understanding is established of what really happens now.

Having mapped the process, the next things the team needs to understand are the true measures of performance of that process, so that they can begin to ask questions of other organizations in order to carry out the first part of the benchmarking exercise, i.e. the external comparison.

8.1.5 Questionnaire construction

In order to gather the data required for the benchmarking study, the team needs to understand the basics of putting together a questionnaire. The different types of questions (closed and open questions), and the usefulness of these different types of questions needs to be understood if the team is to be expected to put the questionnaire together correctly.

Every benchmarking study, and every situation, is different. Therefore, the data collection tool needs to be designed and developed to meet a particular objective. Some processes may require a lot of quantitive data, e.g. an invoicing process requires questions – such as 'how often?' and 'what percentage?' etc. – so a tool would need to be designed to capture sets of hard data. If a human resource process was the subject of the benchmarking study, then questions like 'what is your recruitment policy?' and 'how do you carry out skills evaluation?' require a different type of collection tool.

Once the information needs have been mapped, other things need to be taken into consideration. The data collection exercise is only valuable if it yields the information needed. This may sound obvious, but experience shows that it is often easier to collect less than useful information than that which is really important to the study. The team needs to have a clear idea of:

- how the data analysis will be conducted;
- how the report is to be presented; and
- how the subsequent action plan will be developed.

The means of collecting has to be capable of helping in these areas and this must be borne in mind when constructing it. A well thought out interview proforma, for example, will help the interviewer manage the time allotted, manage the opportunity and return with the information needed for the study.

Basic hints and tips about questionnaires

- Make sure the questions are clear.
- Provide definitions, particularly for any measurements.
- Try to use terms that are generally understood in the industry.
- Be very careful not to use company jargon.

Each team member should try to get the questions answered internally, before the team looks at the answers collectively. This exercise is valuable in that it helps check for inconsistencies. If it proves difficult to answer some of the questions, the team may decide to modify the question or modify the measurements selected. A well documented, flowcharted process, one which relies on hard data will obviously minimize the levels of inconsistency.

If the process is really understood by the team putting together the questionnaire, then good questions may literally fall off the flowchart. This is fine if the purpose of the benchmarking study is gap closure.

However, where the benchmarking is to assist in the radical redesign of processes then a less structured tool, which allows learning to take place, is required. If it is accepted that the current process is just no good, limiting questions to it may well result in missed opportunities. Here the tool may need to ask questions such as:

- Is there anything else that you do?
- Can you tell me about . . . ?
- How many stages?
- How is that managed?

Here the purpose is simply to learn as much as possible, so that after a few visits a clear idea of possible solutions has emerged.

From a pioneering standpoint, when not much is known about the subject area, if little has been published on it, not many people have done work on it, and so on, then the data collection tool may be very different. The tool would be deliberately designed as an opportunity for learning. Questions are less structured and the brief of the team is to learn, investigate and interrogate.

8.2 Data collection methods

There are many different types of data collection methods, from telephone surveys to site visits and interviews. The team has to have consulting skills; it has to know how to gain entry into an organization, understand the best methods of doing so, and understand the need to build up a rapport.

8.2.1 Ways of collecting data

Internal

The service engineers, salespeople, and other people who have worked for direct competition, suppliers, etc. are but a few of the many potential sources of valuable information available without stepping outside your own front door.

Public domain

Sources include newspapers, magazines, trade publications, trade shows, advertisements, company reports, patent records, university and business school research papers, professional bodies, organizations like the European Foundation for Quality Management (EFQM), etc.

Some of these will provide not only subject matter information, but also names of possible future contacts. People within your organization who are members of professional and industry associations could well have ready access to completed research.

User groups also conduct surveys among members and share information on a regular basis. Although the focus of a user group is often fairly narrow, it can provide a wealth of data on work processes, business practices, perception of different vendors, and so on.

Mail surveys

Although many, relatively complex questions can be asked, respondents can complete at leisure, and large numbers of companies can be surveyed with very little cost, there are many disadvantages to this means of gathering data. Some of these are:

- it is not easy for respondents to obtain clarification of any aspect of the survey;
- the sample returned may not be representative of the population surveyed; and
- the response rate to mail surveys is frequently very low, so further time may be required in follow up.

Personal interviews
This is the best means of understanding data as it is collected. Responses can be probed there and then. It is also very effective for gathering information on business practices. It is sometimes difficult to arrange and can become expensive if more than a few companies are involved.

Telephone interviews
This method has a good speed of response, the ability to ask for clarification, and relatively low cost. It does, however, prohibit the use of any visual aids and time is obviously very limited.

Group interviews
This is like a multiple personal interview and therefore requires a very skilled facilitator, due to the highly interactive nature of the exercise. It is very difficult to arrange and without a great deal of discipline from the interviewer, data which is irrelevant to the study can easily pollute the outcome.

Reverse engineering
Purchasing competitive products and systematically taking them apart can be extremely useful for benchmarking product cost, quality and design practices. It obviously requires skilled technical personnel.

Quotations
This is used frequently to determine competitive pricing and practices and is a good way to benchmark cost. However, care needs to be taken about the validity of the quote as any quoting could be deliberately low to win your business. Overuse of this method could easily result in a company not quoting or, worse, provide you with meaningless quotes.

Tours of other organizations' premises, factories
The nature of the exercise exposes the visitors to many different activities at the same time. Therefore, whilst good insights into business practices can be obtained it is vital to select the right people for the visit and make sure individual roles and responsibilities are understood beforehand.

The surrogate site visit
There is nowadays much secondary information available which can provide important information about a particular site without having to

visit in person. For example, some organizations produce videos of particular operations. Studying this information could well obviate the need for a long distance, expensive site visit.

8.3 Which method(s) to use?

Asking the following questions will help determine the most appropriate method(s) of data collection.

- How quickly must the project be completed?
- How complex is the information to be gathered?
- How important is the level of accuracy?
- Is hard data required or will a general trend suffice?
- Are any special skills required available?
- How much clarification is likely to be required?
- How much resource is available for this project and for how long?

In making your decision, you may find that you need to use more than one method. Some common combinations are:

- mail survey followed by telephone interview;
- initial telephone contact to establish any level of interest followed by mail survey then telephone follow up;
- personal interview then telephone follow up;
- initial telephone contact to establish any level of interest followed by personal interview; and
- plant tour followed by personal interview.

A question which now arises is: who is going to collect the data? There are several alternatives.

1. *Yourself* You know what you want and you understand your subject matter, so if you gather the data yourself you will not have to worry about 'buying' into the results. On the other hand you may not be objective and could therefore screen out information.

   Additionally, you may not personally have the time nor the necessary skills required for the particular data gathering method required by your project.
2. *Other members of the benchmarking study team* If there are members of the team who have the necessary skills and a knowledge of the subject matter then, providing time allows, they should be considered.
3. *Another department* There may be individuals in other departments or organizations who already have contacts within the benchmarking partner company.

   However, they may be unfamiliar with the subject matter, inexperienced in the use of data gathering methods or have no 'buy in' to the project and therefore be unwilling to undertake the task.

4. *Consultants* Most consultants have expertise and contacts in specialized areas, can act as an objective third party, have access to industry databases and studies, can handle complex projects and are familiar with research and analytical techniques.

   However, they can be extremely expensive. Great care is needed in deciding to use a consultant instead of internal people who are close to the subject matter.

Ultimately, the requirements of the project should determine the best means of collecting the necessary information.

There are also legal and ethical points to take into consideration before starting to collect data. It is important to remember that benchmarking is an open, honest, above board sharing of information. Therefore:

- Misrepresentation is out! Do not, under any circumstances, make false representation about the purpose of your data gathering exercise.
- Do not divulge proprietary data to anyone.
- Do not entice suppliers by promising business in exchange for information.
- Do not, under any circumstances, try to acquire data on proprietary products or processes.

The benchmarking study team should now be in reasonable shape to begin the data collection exercise.

9

The translation of benchmarking to business results

Benchmarking, as an unfocused exercise, wastes human resource, time and money and does nothing to help business results. This may sound obvious but there are many organizations who are doing just that. Because 'benchmarking' is currently in vogue there seems to be an urgent need to be 'doing' some of it.

The **quadrant model** (Fig. 9.1) clearly shows that benchmarking is more than just a comparison between two or more organizations. It shows that a study begins (top left) with the question 'What should we benchmark?' This really means: 'What area of the business is delivering less than the performance level required?' It calls for a detailed understanding (bottom left) of how results are achieved – what processes, practices, methods, etc. deliver what results? This needs to be clearly understood before any worthwhile comparison can be made.

Therefore, to be used to maximum effect, benchmarking exercises need to be explicity linked to the achievement of the key business priorities. This means having a clear understanding of:

- what the organization is trying to achieve;
- where the organization is against that goal;
- critical success factors (CSFs);
- the key business process which has the strongest correlation with the CSFs; and
- the relationship between processes (enablers) and results.

This basic understanding now forms the basis of many worldwide quality award models, which recognize the important role benchmarking can play in driving actions which lead to superior performance.

As noted before, the most well known quality awards are:

- the Deming Prize (Japan);
- the Malcolm Baldrige National Quality Award (MBNQA) (US); and
- the European Quality Award (launched in 1992).

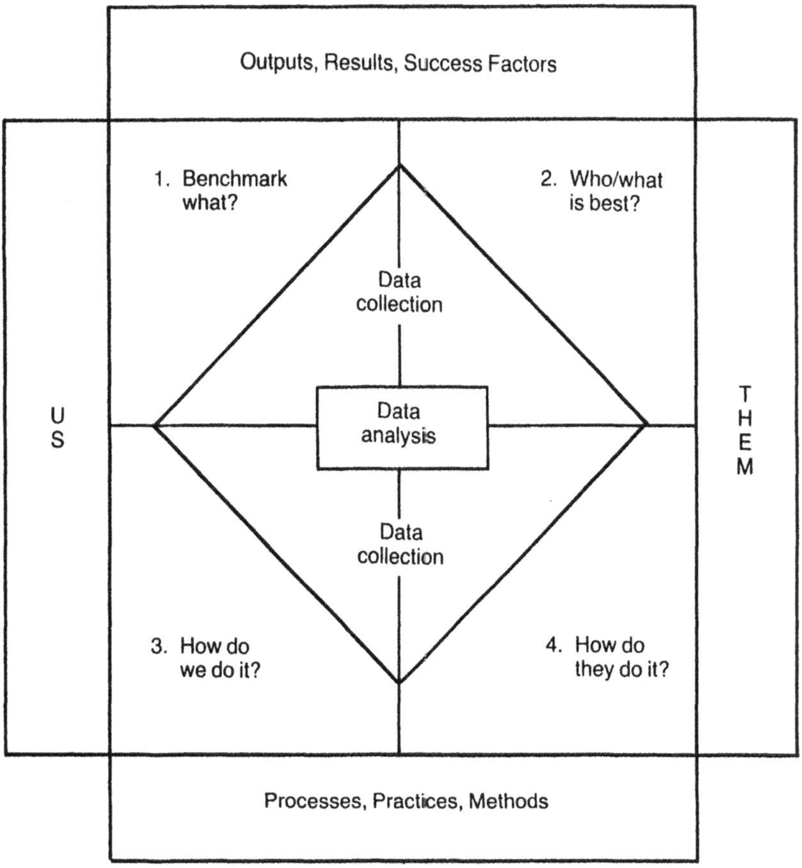

Figure 9.1 The benchmarking quadrant.

The Malcolm Baldrige National Quality Award (MBNQA) in the US requires explicit evidence of the use of benchmarking in its submission document as does the European Foundation for Quality Management (EFQM) with its European Quality Award (EQA).

The MBNQA is broken down into seven areas, some addressing enablers, and others, results:

- leadership;
- information and analysis;
- strategic quality planning;
- human resource development and management;
- management of process quality;
- quality and operational results; and
- customer focus and satisfaction.

The Baldrige model has four basic elements.

1. **Driver** Senior executive leadership creates the values, goals, and systems, and guides the sustained pursuit of customer value and company performance improvement.
2. **System** System comprises the set of well-defined and well-designed processes for meeting the company's customer, quality and performance requirements.
3. **Measures of progress** Measures of progress provide a results-orientated basis for channeling actions to delivering ever-improving customer value and company performance.
4. **Goal** The basic aim of the quality process is the delivery of ever improving value to customers.

Each of the seven criteria categories shown in Fig. 9.2 are subdivided into:

- *Examination items* There are a total of 28 examination items in the seven examination categories. Each item focuses on a major quality system requirement. All information submitted by applicants is in response to the item requirements.
- *Areas to address* Each examination item includes a set of areas to address (areas). The areas serve to illustrate and clarify the intent of the items and to place emphasis on the types and amounts of information the applicant should provide.

  In one particular area, applicants are invited to:

  > describe the company's processes . . . and uses of competitive comparisons and benchmarking information and data to support improvement of quality and overall company operational performance.

The European Quality Award has nine areas, again split between 'enablers' and 'results', as shown in Fig. 9.3. As with the Baldrige model, a great deal of emphasis is placed on demonstrating that benchmarking has an important part to play in driving the company to achieving superior performance.

There are many quality awards in the world today since the launch of the Deming Prize in Japan in the 1970s, including:

- the National Aeronautical & Space Administration (NASA) George M. Low Trophy presented annually to suppliers and contractors who 'have demonstrated sustained excellence, customer orientation and outstanding achievements in a total quality management (TQM) environment'; and
- the Australian Quality Award which 'acknowledges outstanding achievements in organization wide implementation of the quality

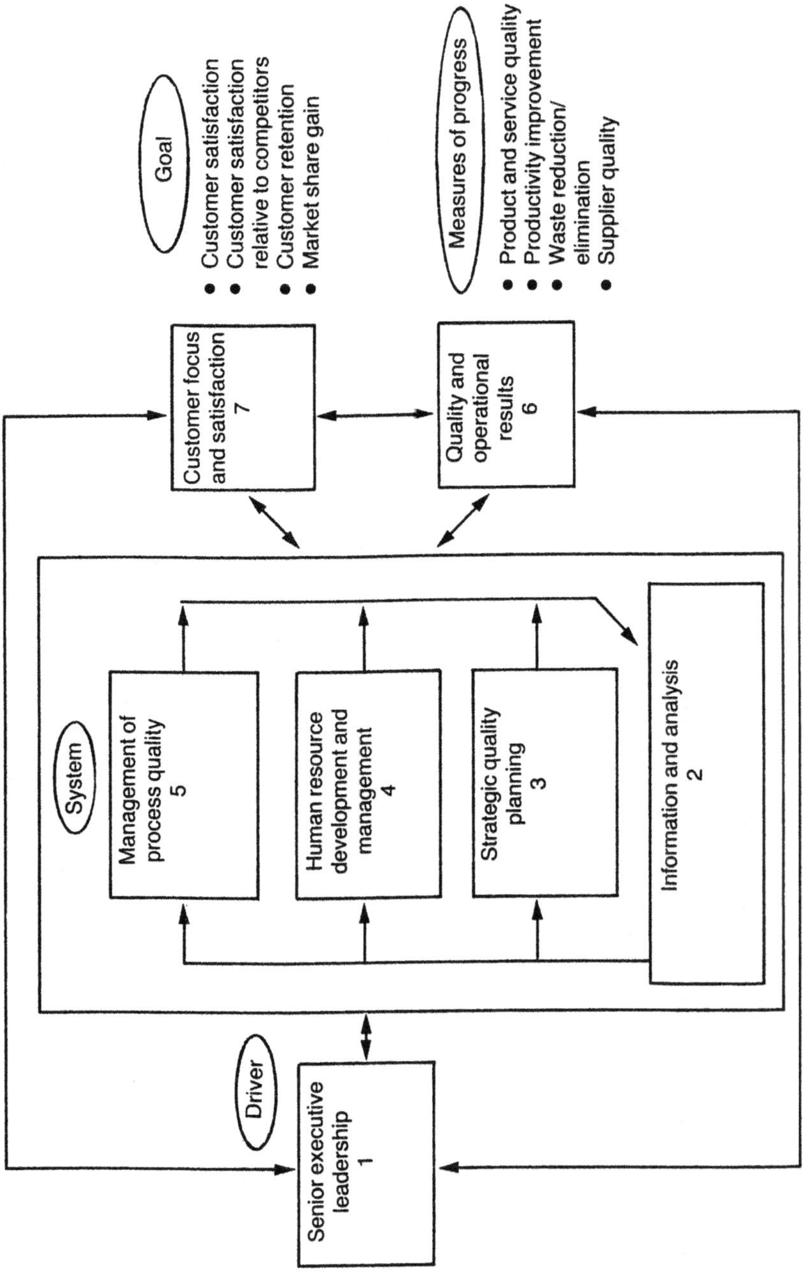

Figure 9.2 Baldrige award criteria framework – dynamic relationships.

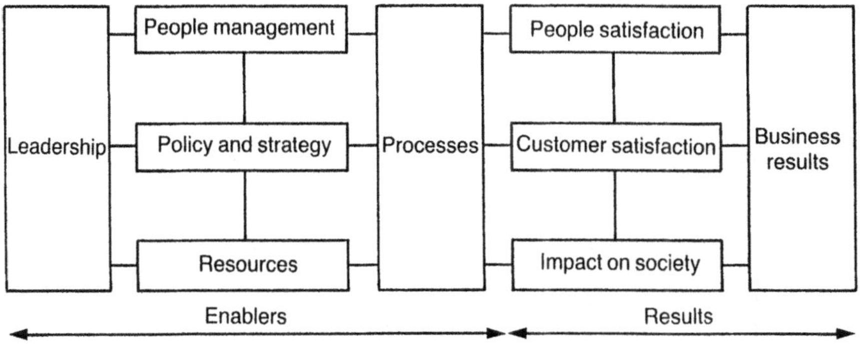

Figure 9.3 The European Foundation for Quality Management model.

culture and its direct link to productivity and international competitiveness'.

All these awards have two things in common:

1. they all require explicit demonstration of the use of benchmarking; and
2. they help an organization take a snapshot of its current position against a set of criteria, usually shown as a percentage score against an 'ideal' or as a score out of 1000.

In 1992, after a long, hard look at its total quality strategy Rank Xerox Limited launched its own self-assessment model called Business Excellence Certification. In this model, described briefly in the company's submission document for the inaugural 1992 European Quality Award, each operating unit is asked to rate its performance in six major business categories as shown in Fig. 9.4.

These six categories are broken down into more than 40 different items and each one is scored against its desired state on a scale of 1–7. The scoring system itself is multidimensional, that is to say it looks at results, the approach taken to achieve the results and the pervasiveness of the approach. So the notion of linking results to enablers is taken down to this micro level. The resulting information is reported within the unit itself and within Rank Xerox as a whole and is used as a guide for improvement efforts.

The self assessment is followed by a validation step, which is a simulated site visit examination with guidance and coaching from qualified examiners – all of whom are senior Rank Xerox line managers from other units.

Some three months later comes the certification examination, carried out by different examiners. If the unit meets the necessary criteria, it is certified. The process of 'assess, learn, plan and improve' takes one year. The analysis of the results forms the basis of the certification criteria for the following year.

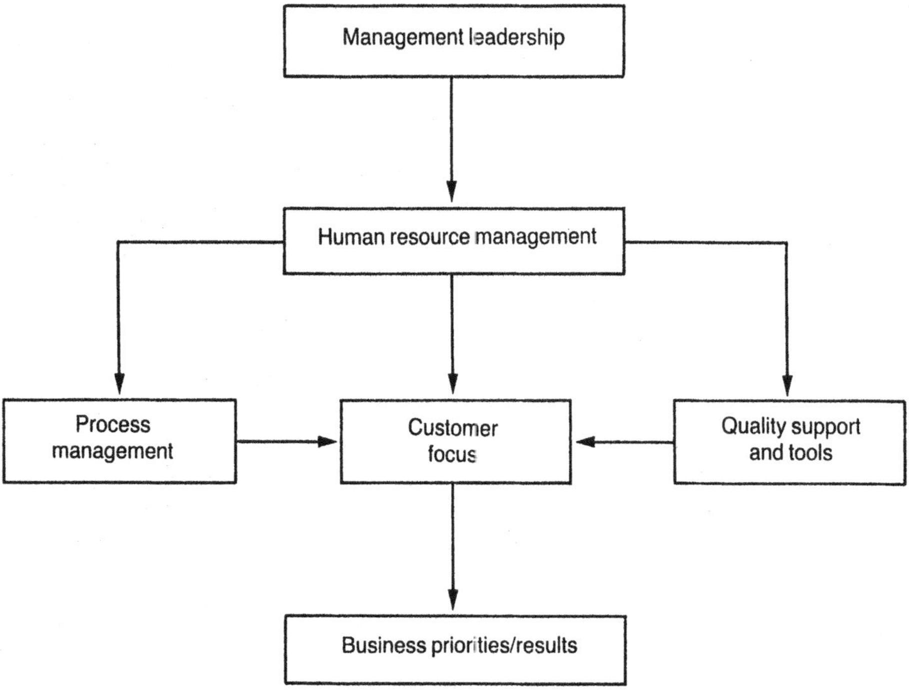

Figure 9.4 The Rank Xerox business excellence certification model.

It has also proved extremely useful in helping the company to improve even more the way in which it uses benchmarking, by identifying one of two systemic areas for improvement. These have since become the targets for company-wide benchmarking studies to achieve superior performance on a company basis. By sharing the self-assessment results on a company-wide basis, it has helped identify areas where improvement is required at a local level, thus increasing the consistency of the use of internal benchmarking.

The use of models like these really help to focus the quality effort and the business effort in the same direction. All too often these efforts, far from supporting each other, pull against each other, to the ultimate detriment of the business itself. The two most common reasons for the failure of 'quality programmes' are:

- not being focused on the achievement business results; and
- not being integrated into the business planning cycle.

These models can be, therefore, very powerful when used to overcome obstacles like these. Overlay that with the use of benchmarking to strive continually for superiority, policy deployment to make sure that critical actions are cascaded in the organization in such a way that everyone understands his/her part and suddenly world class performance is much nearer.

10

The link between benchmarking and performance measurement

10.1 Introduction

Benchmarking is not measurement itself but a process of establishing gaps in performance and as such ensuring that an action plan is put in place to close identified gaps and measured to verify. As Fig. 10.1 illustrates, continuous improvement, although a positive process, may have limitations in so far as the improvements achieved may not be adequate for establishing high competitive standards. The process of benchmarking, however, ensures that continuous improvement is seen externally in terms of the competitive standards it achieves.

Performance measurement, if only internally focused, may have great limitations because it could be considered as focusing on effectiveness rather than competitiveness. Benchmarking ensures that performance establishes competitiveness and best practice through doing the right things, right first time in the eyes of the end customer.

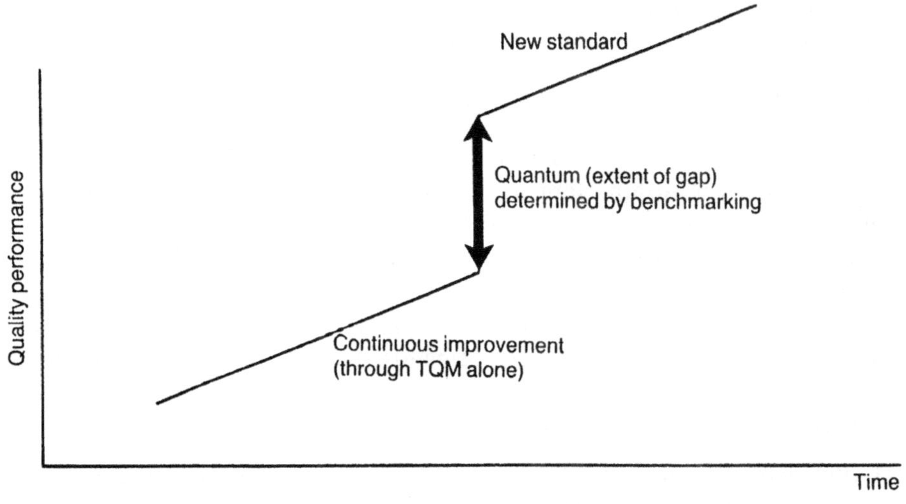

Figure 10.1 Benchmarking: a means for bridging performance gap.

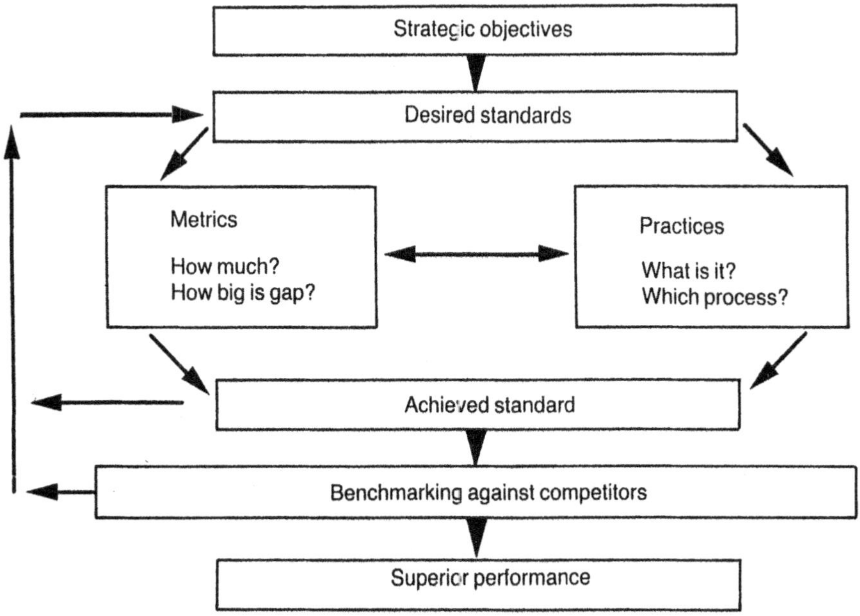

Figure 10.2 The link between benchmarking and performance measurement.

Figure 10.2 illustrates the link between benchmarking and performance measurement.

Benchmarking leads to action and eliminates complacency, as Camp (1993) argues:

> Benchmark findings and operational principles based on them must be converted to action. They must be converted to specific implementation actions, and a periodic measurement and assessment of achievement must be put in place.

10.2 Understanding performance measurement

To understand how benchmarking goes hand in hand with performance measurement, it is important to discuss the recent evolutions in performance measurement. These were triggered by a new need to measure non-financially, in value-added terms, caused by the limitations of traditional standard costing techniques.

There is wide recognition that cost-based measurement systems are inadequate for modern business demands, and in many instances they present a major obstacle to real business progress and improvement.

Cost accounting systems are:

1. inconsistent with new methods which place emphasis on quality, time and value for the end customer;

2. inhibiting organizations from responding to changes in the marketplace flexibly by creating rigid structures;
3. unable to take into account the fact that competitive parameters in the 1990s are not just cost based – customers pay attention to other aspects such as delivery, time, quality; and
4. not encouraging the development of a long-term culture based on continuous improvement.

As Dixon *et al.* (1990) argue:

> The typical approach to improving performance measurement systems has been to focus on improving the cost accounting system. However, the real question is: How can firms measure performance in ways that foster competitive improvement?

On the other hand, Kaplan (1990) reports on the following frustrations that managers feel when using traditional costing techniques in a modern business context:

- measures producing irrelevant or misleading information impede progress and competitive advancement;
- measures conducted in isolation do not give senior management the full picture for making effective decisions on strategy;
- traditional performance measures do not take the customer's perspective either internally or externally; and
- bottom line performance measures are historical measures, thus making it difficult for corrective action to take place.

Kaplan (1990) argues that:

> traditional summary measures of local performance – purchase price variances, direct labour and machine efficiencies, ratios of indirect to direct labour, absorption ratios, and volume variances – are harmful and probably should be eliminated, since they conflict with attempts to improve quality, reduce inventories and throughput times, and increase flexibility. Second, direct measurement is needed of quality, process times, delivery performance, and any other operating performance criterion that companies want to improve . . . Third, financial summaries should exploit available information processing technologies to capture the actual consumption of resources by individual batches or processing lines.

Johnson (1990), another guru of management accounting, supports concerns expressed by various writers. He argues that businesses grew to rely on historical data to make decisions. However, in a modern business context, traditional performance measures are incompatible and inhibitive. He concludes:

> Managing with information from financial accounting systems impedes performance because traditional cost accounting data do not track sources of competitiveness such as quality, flexibility, dependability, and service in the global economy. At best, costs are imperfect signals that a problem exists; they provide no clues as to what the problem is, or how to treat it.

10.3 Best practice in financial measurement – what happens in Japan?

Japanese superiority has often been attributed to their dedication and obsession with quality and a continuous focus on the customer. One area, however, which is less talked about is the use of management accounting systems for controlling and managing various processes.

Japanese management accounting systems are often found to support a commitment to customer-focused, process-oriented and innovation-driven activities. Management accounting systems are an integral part of corporate strategy and are designed to help deliver various strategic goals in manufacturing, marketing and other key functions.

The following are reasons for using management accounting systems in Japan (Hiromoto, 1993):

- to motivate employees and focus their energies so long-term strategies are delivered;
- management accounting systems are certainly not used for the purpose of providing senior management with precise data on costs, variances and profits;
- accounting plays an influencing role rather than an informing one;
- accounting practices in Japan reflect a corporate commitment to delivering market driven strategies rather than to maximize short-term benefits;
- workings of management accounting techniques are based on establishing target costs based on estimates of competitive market prices – it is then left to managers to set benchmarks for achieving the target cost objectives;
- early stages of innovation activity are crucial for achieving target costs – once the product/service is at the production stage it becomes very difficult to economize a great deal;
- unlike standard cost systems which reflect an engineering, and technology-driven management whose goal is to minimize variances between budgeted and actual costs, Japanese practices are market-driven, place more emphasis on ensuring that what is produced is wanted by the market and will perform successfully. As Hiromoto (1993) explains:

How efficiently a company should be able to build a product is less important to the Japanese than how efficiently it must be able to build it for maximum marketplace success.

- accounting systems in Japan are used to help create competitive opportunities for the future rather than quantify what has already been achieved; and
- performance measurement in Japan concentrates more on quality, speed and cost. Placing more emphasis on non-financial criteria for measuring performance ensures that employees become motivated to focus on other criteria rather than become distracted by an obsession with cost.

10.3.1 Management accounting at Daihatsu Motor Company: an example of best practice in Japan

This is one of the best examples of market-driven companies in Japan. It uses the *Genka Kikaku* (Hiromoto,1993) product development system in all its factories. This process is owned by the product manager who is also the project leader from design of the new car all the way to sales. Through involvement of all the various functions, the product manager gathers information on capability and ideas about the type of specifications and features the new car should have; he then makes recommendations to senior management for issuing a new product development order.

Daihatsu then establishes a target selling price based on the market and determines a target profit margin in line with its strategic plans and financial projects. The target costs are arrived at by comparing standard costs achieved without any new innovation, using existing capability and a cost estimate based on new investment. By taking a middle ground between the two costs, the adjusted price-profit margin cost becomes everybody's target.

Design is optimized to meet the target cost with an iterative process of calculating estimated costs using various techniques such as value analysis and value engineering. At the product stage a similar process is used to ensure that target costs are met and costs are well managed. Techniques used include total plant cost management and per unit cost management (*dai-atari Kanri*).

Reports from the *dai-atari Kanri* system are used for informing senior managers but also to inform workers on costs which can be driven down by continuous improvement, i.e. controllable costs. The target costs set in the first place will no longer represent the benchmark once a model is in full production. After one year the benchmark becomes the base line and a new target is set and so on, until capability is used to the optimum (Hiromoto, 1993).

10.4 Integrating performance measurement with benchmarking

The above example illustrates very well how benchmarking can represent the trigger for performance measurement, and how it can be an integral part of it. As Fig. 10.3 illustrates, the dynamics of the competitive wheel can be represented by various key elements.

- *Planning/setting standards of performance*
  Similar to the Diahatsu example, standards of performance need to be set, based on existing knowledge of capability.
- *Performing and measuring*
  Establishing the actual standard and comparing with the set standard – this is the start of a process which attempts to close gaps and conduct improvements.
- *Conducting improvements/closing the gap*
  Through benchmarking and a process of examining practices and measures, new methods/practices can be used to close performance gaps and carry out necessary improvements.

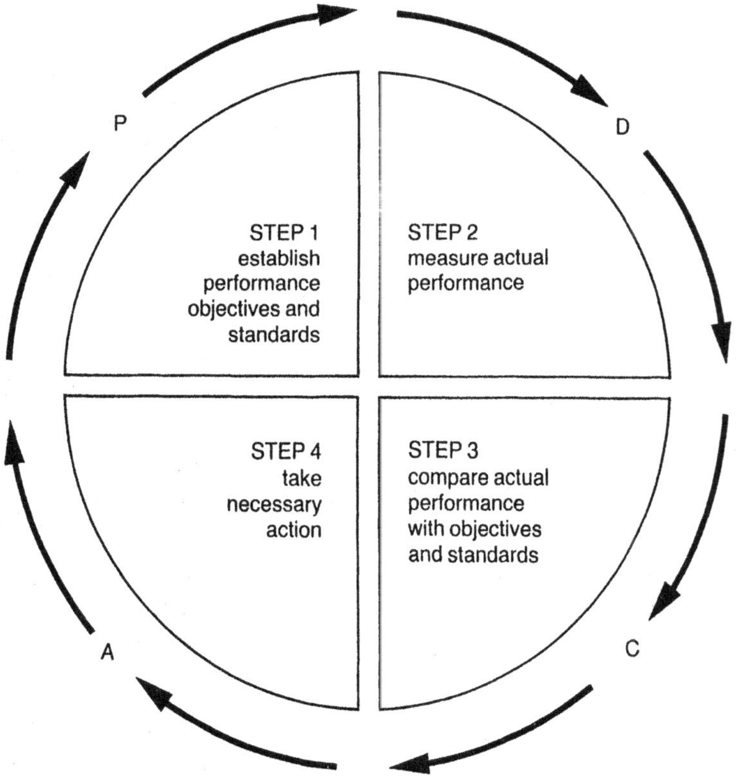

Figure 10.3 The process of continuous improvement.

- *Revise existing standard/set new one*
  Ultimate action should lead to a continuous review of performance and a re-setting of targets so that capability is optimized all the time by delivering quality to the end customer, supported by an aggressive effort of keeping costs down.

The two processes of benchmarking and performance measurement work hand in hand:

- benchmarking is a process which 'sets the bar' at new heights knowing what levels of competition are;
- performance measurement is a process of 'fitness' – preparation, rehearsal, continuous training to ensure that the heights set are cleared.

Figure 10.3 illustrates how the wheel of improvement is carried out and how efforts should be conducted in order to close competitive gaps. Figure 10.4 on the other hand places emphasis on any likely action which needs to take place so that gaps are closed.

Three possible outcomes are likely to happen as a result of measurement:

- performance of actual is less than standard set and therefore action needs to take place;

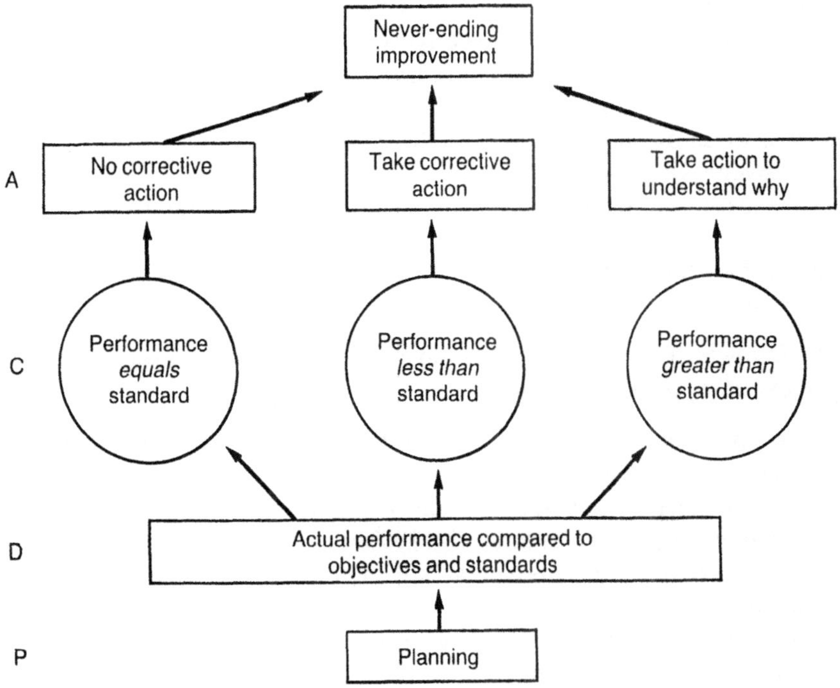

Figure 10.4 Improvement through action taking.

- performance of actual is equal to standard, therefore no action is necessary; or
- performance of actual is higher than standard, therefore action needs to take place to understand why.

10.5 Deploying measurement throughout the organization

Integrated benchmarking/measurement should be geared towards helping businesses achieve desired goals, by translation total effectiveness internally to superior performance externally. The integrated measurement effort has to be:

- continuously improved;
- sustainable;
- continuously focusing on the customer;
- continuously building internal capability; and
- continuously feeding in new knowledge and expertise.

10.5.1 Critical factors in performance measurement deployment

There are two key factors which need to be considered for effective deployment of performance measurement.

1. Performance measurement is driven by business strategy. The process is established through: vision → CSFs → strategy → performance measures → indicators/trends/targets → setting standard and review;
2. Measures must be designed to reflect the following:
   - a focus on the customer;
   - a focus on outputs/value-added activity (time, speed, cost, responsiveness); and
   - short-term, to encourage continuous learning and innovation.

10.5.2 The deployment of measurement in the supply chain: an example

The following is an example of deploying performance measurement in the context of supply chain management.

1. **Vision/mission**: to build a competitive advantage through provision of superior service to our customers.
2. **Critical success factors**: to deliver to customers all order lines in completeness right first time with 100% delivery reliability.
3. **Prime performance measures**: key measures for each core process within the supply chain.
4. **Secondary measures**: key measures of value added activity in each core process within the supply chain.

Figure 10.5 illustrates the deployment of measurement in the supply chain.

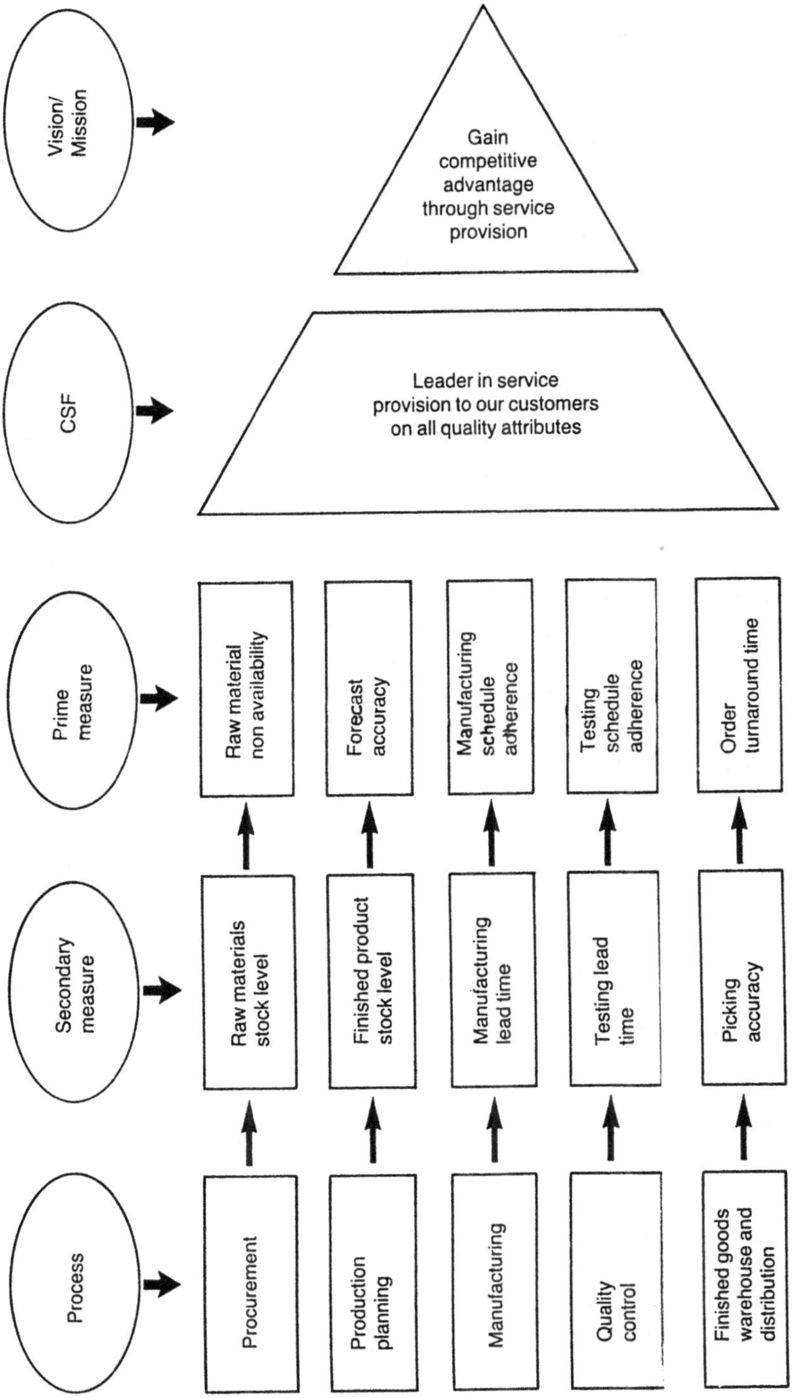

Figure 10.5 The deployment of performance measurement in supply chain.

10.5.3 The deployment of performance measurement at General Electric (GE)

GE is a global company manufacturing industrial and consumer products. A new performance measurement system (Johnson, 1990) was introduced in 1987 in a product assembly line based on activity management principles. Figure 10.6 illustrates the performance measurement system introduced, which was based on the following set of conditions:

- interdependency exists between critical success factors and operating performance measures for the whole of the product assembly line;
- improvements internally at operating level within the assembly line should lead to improvements externally to the customer;
- as non value-added activity is identified and eliminated this should automatically lead to improvements at the operational level of all the assembly line.

Figure 10.6 illustrates how at GE business measures run parallel to customer-based measures driven by value-added activity. There is a sharp contrast between the new measures and old ones which were product-based, internally focused on activity rather than externally focused on customer service, and which focus on individuals rather than the process.

10.6 An integral approach to measurement

A useful model of performance measurement (Kaplan and Norton, 1993) is based on a 12-month research project on 12 leading companies. The model is referred to as the 'balanced scoreboard', and covers all areas critical to business competitiveness. It addresses four sets of issues.

1. **Customer perspective:** the premise is that all businesses exist to satisfy customer requirements. To succeed they have to start with the customer and not with the needs of the organization itself. Measurement has to be externally focused, with external data including service, quality, responsiveness and cost.
2. **Process capability:** building capability internally is essential to becoming competitive. Optimization of process performance in terms of quality, speed, delivery, cost and synergy through cross-functional activity and teamwork is essential.
3. **Focus on innovation:** modern competitiveness is based on fulfilling customer requirements through creativity and innovation. Building a learning organization where people productivity is the main focus, the belief in learning through improvement is essential. The consideration of people as the main asset is crucial and measurement of employee satisfaction and employee attitudes is crucial.
4. **The financial perspective focusing on the shareholder:** shareholders are another set of customers and value added to shareholders must be

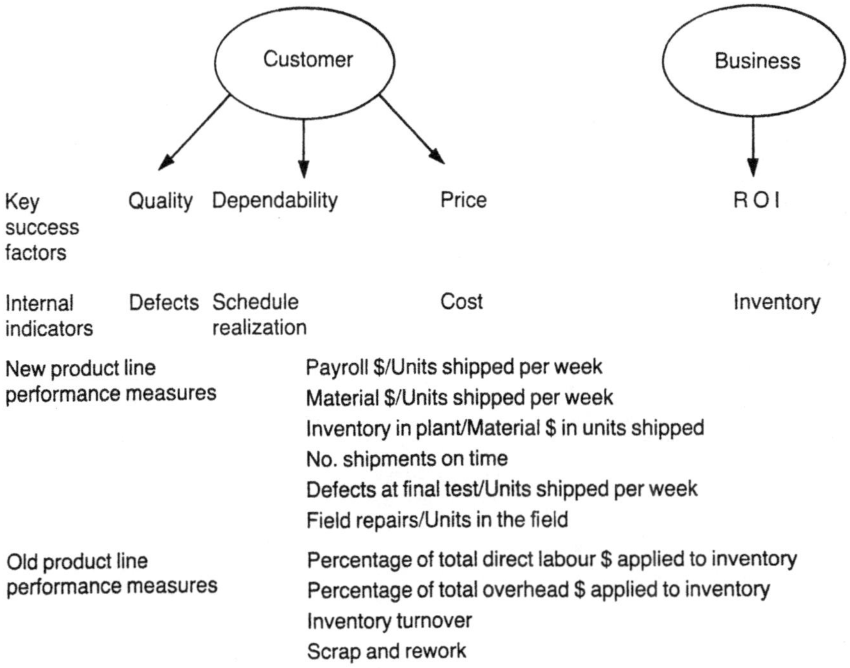

Figure 10.6 The deployment of performance measurement at General Electric (GE).

tracked continuously. Delivering performance which would satisfy shareholders has to be looked at in an overall context of corporate business objectives. Sound strategies tend to blend short-term and long-term impact with the view of optimizing delivery to customers, shareholders and the organization itself.

Figures 10.7 and 10.8 illustrate the balanced score card model and an example of its application, respectively. The usefulness of this model lies in its ability to put strategy and vision first rather than control. It supports:

- teamwork;
- empowerment;
- continuous improvement;
- the implementation of quality principles;
- the use of tools and techniques for improvement; and
- innovation through continuous improvement and a focus on the end customer.

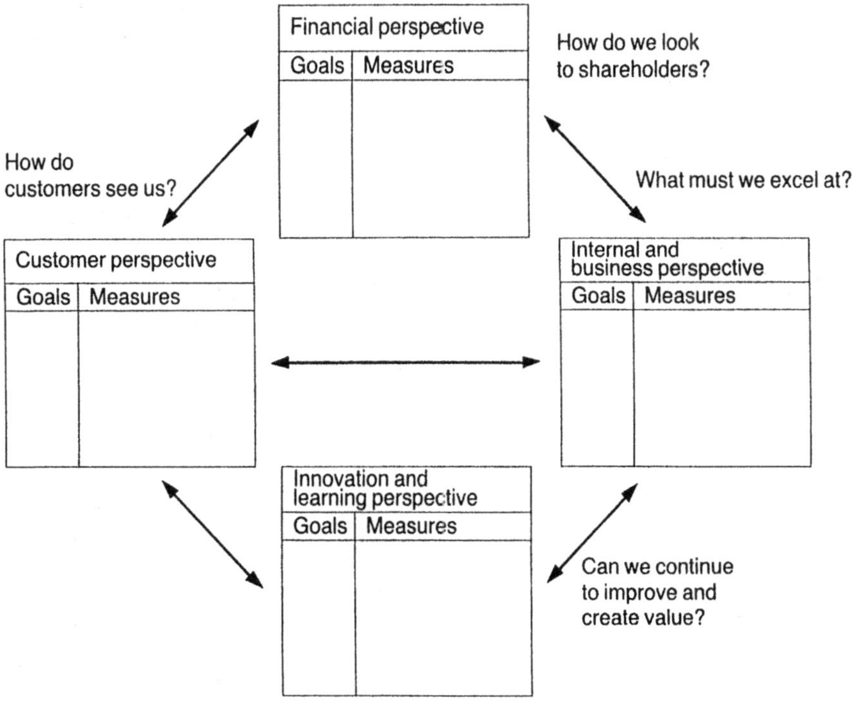

Figure 10.7 The balanced scorecard model: an integrated performance measurement system.

10.7 An integral approach to benchmarking

Integrating performance measurement with benchmarking can be done in two stages:

1. *Effectiveness stage* Seven steps can be used in this first stage. These include understanding all internal processes through simple techniques of total quality management, such as flow charting or process mapping, establishing the performance levels of each process, identifying areas for improvements, conducting the improvements, measuring time and time again, and setting high standards of consistency and controlling and managing each individual process using the plan–do–check–act (PDCA) approach;
2. *Competitiveness stage* Nine steps can be used in this stage. It is very important first of all to select each individual process carefully for high suitability for benchmarking. The strategy then recommends a careful choice of partners and agreement of a benchmarking strategy conducting the exercise, acting upon the outcomes of the exercise, comparing standards and repeating the experience time and time again using the

Financial perspective	
Goals	Measures
Survive	Cash flow
Succeed	Quarterly sales growth and operating income by division
Prosper	Increased market share and ROE

Customer perspective	
Goals	Measures
New products	Percentage of sales from new products Percentage of sales from proprietary products
Responsive supply	On time delivery (defined by customer)
Preferred supplier	Share of key accounts purchases Ranking by key accounts
Customer partnership	Number of co-operative engineering efforts

Internal and business perspective	
Goals	Measures
Technology capability	Manufacturing geometry versus competition
Manufacturing excellence	Cycle time Unit cost Yield
Design productivity	Silicon efficiency Engineering efficiency
New product introduction	Actual introduction schedule versus plan

Innovation and learning perspective	
Goals	Measures
Technology leadership	Time to develop next generation
Manufacturing learning	Process time to maturity
Product focus	Percentage of products that equal 80% sales
Time to market	New product introduction versus competition

Figure 10.8 **An applied example of the balanced scoreboard model in electronics.**

PDCA cycle with either existing practices or new ones. The process of benchmarking can then be extended to all the other processes.

10.7.1 Maintaining the effectiveness of the measurement–benchmarking blend

The following areas have to be addressed if an integrated, effective approach of measurement and benchmarking is to be achieved:

- establishing process capability and maintaining the standards achieved;
- establishing understanding of internal and external customer requirements;
- creating a process for monitoring customer satisfaction levels;
- creating a learning organization and a focus on continuous learning through new knowledge and creativity;
- being aware of competitor moves and always able to influence market behaviour;

- measuring what is important to the customer, i.e. quality, cost, time, innovation;
- looking towards some end objectives such as:
  - achieving customer loyalty;
  - customer retention;
  - employee satisfaction and retention;
  - increased sales and market share; and
  - improved profitability.

In summary, performance measurement and benchmarking are two essential improvements for ensuring that TQM implementation will succeed. TQM implementation seeks to establish a culture of continuous improvement through teamwork, using a problem-solving approach based on tools and systems. However, the result of TQM is measuring performance in terms of quality, cost and delivery. These standards can only become competitive through the practice of benchmarking.

11

Choosing partners for effective benchmarking

Companies often begin their benchmarking exercise by asking 'who should we benchmark against?' This demonstrates a complete lack of understanding of the whole question of benchmarking.

Benchmarking is all about comparison of processes, practices and methods used in one's own organization with those of other organizations and therefore has to begin with a detailed understanding of oneself.

The organization needs to understand where it is trying to get to, what it is in business to achieve (vision, mission, strategy). It needs to understand and clearly articulate and communicate its business goals, both strategically and operationally, and its short-term business objectives.

Against that background, the next question is 'How do we, or how are we going to, deliver those results?' In other words, which processes drive which area of the business? The link between the how (process) and the what (results) must be clearly understood. If, after investigation, any business process appears to have no explicit link to a result, then its usefulness and very existence needs questioning.

Until the critical work processes that drive the business results are really understood there is little or no point in attempting a benchmarking exercise.

Many organizations, when asked, will say that they do, of course, understand their business, but when faced with questions like:

'Do you know which are your critical business processes?'
'Are they documented?'
'Are they flowcharted?'
'Are they measured?'
'Are the measures captured?'
'Are trends reviewed regularly?'

then very often a different picture begins to emerge.

If the answer to any of these questions is 'No' or 'I don't know' a lot of work has to be done, almost as a pre-requisite to carrying out a benchmarking study.

The next pre-requisite to carrying out a benchmarking study is to clearly understand where there is a shortfall in any planned results. This, coupled with the knowledge of understanding the critical process(es) which drive these results will help us to focus on the process or subprocess which is likely to become the subject of a benchmarking study.

Only when these previous steps have been completed and the purpose of the proposed benchmarking study has been written, understood and agreed with senior management should we begin to think about potential benchmarking partners.

All this prework is essential if the benchmarking study is to stand any chance of success.

11.1 Beginning the search for benchmarking partners

Who can help you start putting together a list? Ideas can come from customers, people with whom you work in your department, other departments or other divisions in your company which perform similar functions to your own, members of professional or trade associations etc.

Other ways of researching potential benchmarking partners might include looking in trade journals for companies that have received recognition, won industry awards or been awarded patents, etc. This will help you to begin to assemble a list of companies with reputations for achievement and innovation – likely candidates for your study. Annual reports and business directories like those produced by the business information company Dunn & Bradstreet can also be valuable sources of information.

If your initial list is too long, i.e. more than 12 or so organizations then some work at reducing its size is necessary before thinking about making contact.

11.2 List reduction

Try answering the question: 'how reliable is the information I have?' If the information source is at all dubious then that particular organization should come off your list.

If your research shows that a particular company is probably the best in class but is not renowned for friendly collaboration, you could waste time going nowhere. Experience has shown that focusing on friendly companies, where doors are relatively easy to open, much better progress is made in a much shorter timeframe.

Then, the question 'How sure am I that they use processes which will be of use in this benchmarking exercise?' should be validated against the purpose of the benchmarking study, which should be the focal point for all thought and action from now on.

11.3 Basic pitfalls

Things to avoid in the list reduction exercise are:

- removing companies from lists without adequate information, on the whim of members of newly appointed benchmarking study teams;
- not considering companies from other industries since this demonstrates a lack of understanding of the true nature of benchmarking. If the list contains no one from another industry, be suspicious.
  - Have industry leaders been excluded?
  - Have other departments or divisions within the company not been considered?
  - Have searches of public sources been superficial because people do not know where to look.

Ideally this will lead to a list of a dozen or so possible partners. These may or may not include other divisions of your own organization, be companies in a similar industry, or be in an entirely different industry. Experience has shown that truly innovative benchmarking results have been achieved where the latter is the case.

With the list complete there may be a strong temptation to start reaching for the telephone! However, before any initial contact is made, beware.

- They may say 'No', abruptly.
- They may say 'Thanks, but no thanks'.
- They may think you are trying to sell your products.
- They may show enough interest to ask questions.
  - Why have you come to us?
  - Can you explain the rationale?
  - Could you fax a flowcharted copy of the process you are looking at?
  - What's in it for me?
  - How do we handle sensitive information?
- They may even want to meet you.

Are you fully prepared?

11.4 Making initial contact

There is an assumption that benchmarking automatically provides an entry permit and *carte blanche* access to another organization. This is rarely the case. Why should another organization be prepared to provide information without getting something in return? This issue in benchmarking is often neglected, yet the quality of information gained is reliant on the establishment of a partnership of equals. Sales flair is often required to obtain access and maintain the appropriate contact level in a potential benchmarking partner. Naturally there has to be a willingness to reciprocate and this selling point can facilitate the establishment of a relationship.

One proven way of making initial contact is by telephone, ideally to the most senior person in your area of interest. It is well worth the extra time to identify this person prior to making the call, rather than starting with the switchboard operator. This first contact is vital and thorough preparation is required.

If the company is a supplier or a customer of your own organization, it may be useful to approach the account manager or sales representative to act as an intermediary, to help set up a visit or to identify the most appropriate person.

Part of making initial contact means providing the person with a very clear purpose of the benchmarking study. It is important to note that the purpose statement could (if things went wrong) be useful from a legal standpoint at a later date. This statement should be part of the correspondence requesting the benchmarking visit. The initial contact should also stress interest in industry best practices. It is extremely useful at this point to be able to cite an area of known excellence about the other company.

Remember that when you do get through to the right contact that he/she is a busy person so there is a need to be brief and to the point. Listen very carefully to any questions and always answer truthfully. If you do not know the answer to a particular question, say so! It probably means that your preparation has been less than perfect, but if you then find out the answer, the situation is normally recoverable.

When faced with the question, 'What's in it for me?' an answer along the lines of, 'There is nothing obvious now, but if at some stage you embark on a benchmarking study of your own and consider us to be a suitable partner, then please feel free to contact us' usually works. Any problems around potentially sensitive information are usually resolved by agreeing to sign a non-disclosure agreement.

Initial telephone contact is normally followed up by letter. The purpose of the follow-up letter is twofold: to reiterate the purpose of the study and to summarize the outcome of the telephone conversation. Even if another company readily agrees to becoming a benchmarking partner do not assume that any doors have been opened into that organization or that they are really suitable, at this stage, to assist in the current study.

By this time, the list of 12 has probably been whittled down to six or eight.

11.5 The next step

The next step is to answer the question: 'Is your organization really suitable to be a benchmarking partner in this study?' The next contact must be aimed at answering this question. Preparation once again is all important.

If at, or before, the next point of contact, you identify unsuitable partners, or an organization has second thoughts and no longer wishes to continue, it is important, in the interests of building long-term relationships, to note that you may have to agree at the initial contact stage to becoming a potential benchmarking partner yourself at some point in the future. If this is the case, record the details so that any future contact by another organization can be dealt with professionally.

After the intitial contact by telephone and letter it is now desirable to meet face to face.

Once there is agreement to a site visit, it is worth spending time gathering enough information about the company to be able to talk with a reasonable level of background knowledge. This is useful in beginning to build a relationship. Decisions also need to be taken on how many should go, who should go, what specific skills are needed to ensure a successful visit, what will be the role of each visiting team member, what is the desired outcome of the visit, and on the agenda for the visit.

Benchmarking is a two-way exchange of information preferably based on prearranged questions. It should not be used as a one-way opportunity to collect information simply to gain a short-term business advantage. Remember the one basic principle of benchmarking is not to ask questions of another company that your own company would not be willing to answer. You also need to be cautious about sharing information when there is no agreement to treat the material which is shared as confidential. Benchmarking interchanges of information have to be confidential to the individuals of the companies concerned.

It is a time-consuming task to co-ordinate diaries and set up a meeting. An agenda needs to be developed which is clearly understood and the names and job roles of participants need to be shared by both parties beforehand. It is also useful to take to the visit a thumbnail sketch of your own organization as a handout.

Managing the agenda of a benchmarking site visit must be a two-way process and generally the number of people kept to a minimum. The optimum team size is three people with their individual roles determined in advance. Decide who is going to ask questions, who will know when all the questions have been answered, who is going to be the notetaker, etc. If, during the site visit itself, it is difficult or impossible to make notes, it is extremely important that immediately after the visit the team spends enough time to capture everything that occurred. The team should be debriefed as soon as possible after the visit.

At the meeting itself there needs to be an agreed process to achieve the purpose of the visit. If the agenda is hijacked by one of the parties, then the future of any long-term benchmarking alliance is immediately put at risk. The onus is very clearly on the initiator of the benchmarking study to ensure all these activities run smoothly.

Some things to encourage when arranging or participating in a site visit are obvious. Be professional, be honest, courteous, prompt, adhere to the agenda, maintain focus on the benchmarking issues, introduce all attendees and explain their presence, and use language which is universal, not one's own jargon. Talk with your contact in the other company first, to assist in finding a process owner for the study, rather than referring him or her to another person. Offer to share the general findings when the study is completed. Offer to facilitate a future reciprocal visit, conclude the visit on schedule, and thank the benchmarking partner for their time.

Some things to be avoided at all cost are:

- referring to another company or disclosing information about another company whilst on a site visit;
- broadcasting information about another company by whatever means, without their permission;
- asking for information which you are not willing to provide;
- making contact and setting up a visit without preparation;
- providing hard copy of quantified business performance measurements;
- sharing price plan establishment data; and
- sharing any sensitive data – private, personal or registered information on other companies that are, or have been, benchmarking partners.

At this point in the benchmarking study the number of benchmarking partners should be about four of five. Less than this number will give very limited data for comparison purposes and more could lead to collecting too much data.

If any of the benchmarking partners are in direct competition with each other this could well cause either or both to withdraw from the study, so it is as well to get this out into the open as early as possible. This problem tends to go away once sound relationships have been established and there is mutual trust.

In summary, the first visit should be little more than an ice breaker, a 'getting to know you' meeting aimed at beginning to build a rapport and outlining data exchange principles. Do not expect to have all the information you need after only one visit.

After this first visit it is important to send the benchmarking partner a synopsis of the meeting including your interpretation of any data gathered at the visit. This is a very useful exercise which checks the accuracy of your interpretation and begins to involve the partner in the study.

Following the guidelines outlined in this chapter will go a long way towards establishing long-term benchmarking partnerships.

Part Two

12

Benchmarking applications

The examples of benchmarking applications included in this section of the book are not meant as indicative of any sound methodologies but rather they are intended to illustrate the *applicability* of the concept of benchmarking in a wider context.

In most cases benchmarking tends to be used implicitly, by comparing trends, practices and ways of working. Although most organizations covered in this section act on the outcomes of these comparative exercises, the resulting changes, however, are not continuous but represent a one-off type of behaviour.

The various applications covered could be more powerful if the potential is explored in the context of a proper benchmarking methodology based on:

- quality management principles;
- a focus on the end customer;
- performance measurement;
- using the PDCA cycle for continuous improvement;
- focusing on the process rather than just the outcomes; and
- learning through repeating the task, time and time again, with increasing knowledge, expertise and competitive know-how.

Applications covered in this chapter include the following:

- Utilities sector
- Financial services sector
- Innovation
- Human resources
- Customer–consumer satisfaction
- Marketing
- Education
- Healthcare
- R&D
- Environment
- Quality systems
- Public sector
- Multinational/monopoly situation

12.1 Benchmarking in the utilities sector

The utilities sector of industry is faced with many different challenges. Although in recent years this industry has been more geared towards greater efficiencies and becoming more competitive, unlike any other industrial sector, it still places more emphasis on safety first. In addition, this industrial sector is highly regulated and has to meet continuously set standards.

Benchmarking in the nuclear industry is a well established concept, although it may not have been referred to as such in the past. In 1979, the Institute of Nuclear Power Operators (INPO) was established in the US to promote the highest levels of safety and reliability and to promote excellence in the operation of nuclear power plants. Its mission was to promote best practice and to ensure that standards of safety were tightened more and more.

INPO was also established to promote and encourage the exchange of information on good practices and lessons learned from all the nuclear operating companies. It recognized that by improving and refining existing practices, safety standards could be raised and disasters avoided.

INPO encouraged the involvement and participation of personnel from different functions within utility companies, to promote and exchange information and to enhance knowledge levels of all people involved.

Utilities joined INPO from at least 14 different countries, and one could argue that since its creation INPO has helped enhance standards of safety worldwide to an excellent level.

In 1989, the World Association of Nuclear Operators (WANO) was created. Based on similar principles, WANO is used to facilitate the exchange of information between the various nuclear plant operators worldwide, and to encourage comparison, emulation, communication, exchanges, partnerships and joint projects between the various corporate members. The objectives of WANO are as follows:

- operating experience exchange;
- operator to operator exchange;
- plant performance indicators;
- sharing of good practices.

WANO has ten performance indicators used for benchmarking. All the data are compiled by WANO and made accessible to all the corporate members for learning from each other.

In addition to INPO and WANO, various operators in utilities participate actively in the International Atomic Energy Agency (IAEA). The IAEA is an ideal forum for sharing standards of excellence, learning high standards of safety and reliability, and operating nuclear power plants in an economic and efficient manner.

Therefore, benchmarking in utilities is not an alien concept and has been applied for a large number of years. The utilities sector is an interesting one to look at from the point of view of benchmarking, especially in view of recent changes particularly in the UK. The Central Electricity Generating Board (CEGB) used to enjoy a monopoly position in the bulk supply of electricity within England and Wales. Its restructuring has led to the birth of various operators, including Nuclear Electric and National Power, who have to operate independently and compete in the private electricity market.

12.1.1 Benchmarking at Nuclear Electric

Nuclear Electric has used benchmarking to develop stretch objectives and put its vision in place. Nuclear Electric's vision is:

> We will be a quality company, we will make an operating profit before the levy in 1995, and be a key part of the country's energy supply into the twenty-first century

Through benchmarking, Nuclear Electric has established its strategic framework for driving for the desired goals and established performance goals in the following areas:

- manageable costs per unit of production;
- numbers of staff; and
- output per employee.

Nuclear Electric uses benchmarking for both internal and external comparisons.

Internal comparisons
Nuclear Electric uses internal benchmarking in its various power plants at operational level.

1. **Plant evaluation** An evaluation has taken place using a group of peers to identify good practices and compare them against internationally accepted standards. Internal benchmarking within Nuclear Electric is really for identifying areas for improvement in all the activities, particularly those that affect safety and reliability standards.
2. **Investors in people** This is an opportunity for Nuclear Electric to use its human resources more pro-actively and to encourage innovation at all levels to meet business objectives. The investors in people programme initiated by the British government was used at two lead sites within Nuclear Electric, particularly to determine areas for improvement.
3. **Cost management review** This exercise has taken place in order to look at costs in relation to activities and their worth for overall company objectives. It has helped the company to identify opportunities for improvement and to proceed with plans for rationalizations, but it has also enabled the comparison to take place between the various stations and the numbers of staff involved in all the various activities.

External comparisons
Other benchmarking exercises have taken place within Nuclear Electric: benchmarking its operations against other utilities and also in the wide industry community as a whole.

1. **Outage management** The benchmarking exercise looked at Nuclear Electric's average outage time in all its power plants, to identify opportunities for reduction. As a result, in two power plants the outage was reduced from 17 weeks to ten weeks with a further opportunity to reduce this to eight weeks.
2. **Safety case management** This benchmarking exercise scrutinized the whole safety case management system. The exercise is still ongoing and its purpose is to simplify the structure whilst maintaining the essential elements of safety case management.
3. **Health physics** This is a benchmarking exercise with all the utilities in order to establish best practice in radiation protection activities.

Benchmarking within Nuclear Electric has worked. Output has gone up, costs have come down, safety standards have been maintained and even improved:

- output increased by 29%;
- market share increased by 26%;
- costs reduced by 22%; and
- productivity increased by 53%.

At Nuclear Electric the target for 1995 is to make a profit by re-examining manpower, flexibility and utilization and to bring about the necessary changes in a more radical way by re-examining phase costs and focusing on substantial productivity increases. Benchmarking has given Nuclear Electric the confidence to establish stretch objectives and aim to achieve them. Striving for excellence in Nuclear Electric is becoming a way of life and in the words of its Chief Executive Dr R. Hawley: (Redman, 1993)

> In the private electricity market, there is no room for other than the excellent, our competitors will see to that.

12.1.2 The case of Florida Power and Light

Florida Power and Light was established in 1925. It is the fourth largest investor owned electric utility in the US, and certainly the fastest growing in terms of numbers of customer accounts. Based on 1989 figures, its serviced territory spreads to a population of approximately 5.7 million, covering 27 650 square miles which is approximately one half the State of Florida. Florida Power and Light, based on 1989 figures, was employing 15 000 employees, it had facilities including 13 operating plants, seven operating offices, 45 customer service offices, 72 service centres, 397 substations, 53 300 miles of transmission and distribution lines and it had, more importantly, three million customer accounts.

Florida Power and Light started experience with quality back in the 1970s; however, it was not until the early 1980s that quality was introduced systematically at corporate level. Florida Power and Light won the Deming Prize in 1989 by adhering to the following principles.

1. **Customer satisfaction** The need and reasonable expectation of customers must be satisfied.
2. **PDCA (Plan Do Check Act)** Continuous improvement is based on working at the process, planning what to do, carrying it out, checking, acting to prevent error and improving the process and then repeating the PDCA cycle time and time again.
3. **Management by fact** This concept has two aspects to it: first, collecting data objectively then using the refined information to solve problems; and then acting in order to improve quality in all aspects.
4. **Respect for people** This is fundamental to the Florida Power and Light philosophy in that all employees are self motivated and encouraged to use their creative output. Their skills and knowledge are continuously harnessed to ensure that there is optimum utilization. People are valued as an integral part of the organization.

Through quality, Florida Power and Light has achieved tremendous results. Some are not measurable. For example, it has developed a new corporate culture where all employees have a renewed sense of pride, respect and satisfaction. The climate of work is based on multifunctional levels where everybody is given the opportunity to contribute and use their creative talent. Some employees have started to use the continuous improvement approach in their everyday work. Outside opinions of Florida Power and Light have changed significantly for the better. The customer, the media, stockholders, government agencies, politicians, and even their regulators have started to admire the way Florida Power and Light drives its business towards excellence.

Benchmarking at Florida Power and Light

Based on the Rank Xerox methodology, Florida Power and Light developed their own approach to benchmarking. First by selecting 15 other leading electric utilities to benchmark against, Florida Power and Light tried to obtain data on the various functions, particularly data that could be used to compare quality indicators.

The information obtained was useful. In 1985, Florida Power and Light did not measure up at all against the 15 top performers in utilities. In 1988, they moved from 13th to third place and, by 1989, were moving towards second place. Figure 12.1 shows just how good Florida Power and Light were compared with the companies against which they were benchmarked.

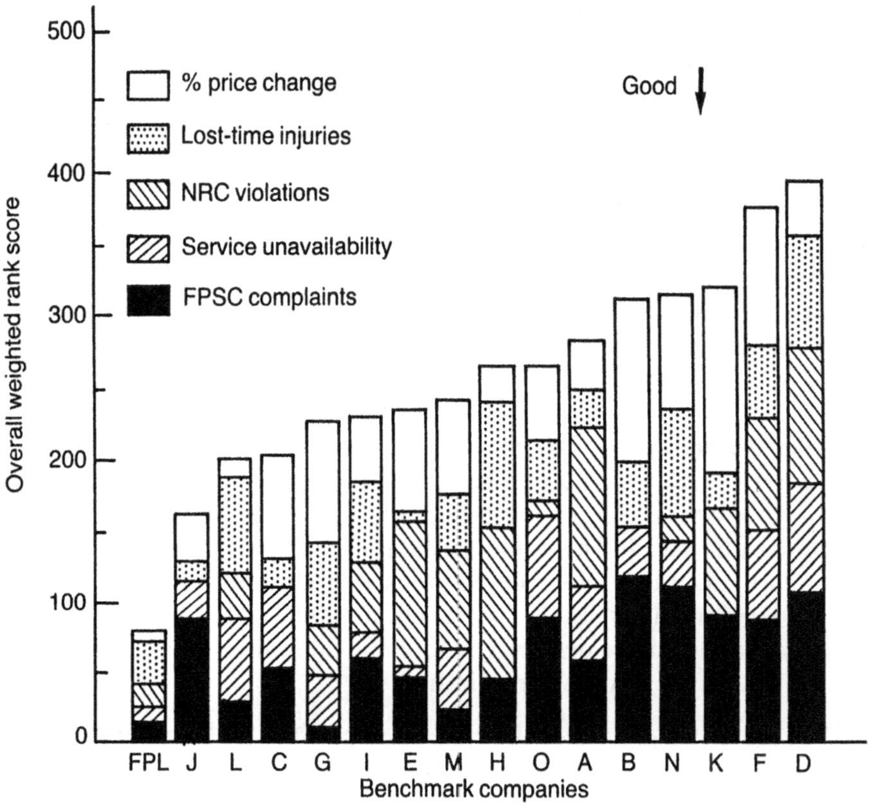

Figure 12.1 Benchmarking in utilities: the case of Florida Power and Light. *Source*: Hudiburgh (1991)

From 1986 to 1989, the percentage of Florida Power and Light customers who reported being extremely satisfied increased from 45% to 55%.

Florida Power and Light learned right from the onset to start with the customer and changed all its processes to focus on the customer continuously.

The customer survey conducted by Florida Power and Light in 1986, to identify programme areas and customer needs, included the following measures:

- accurate answers;
- timely action;
- accurate bills;
- considerate customer service;
- customer programmes and services;
- continuity of service; and
- understandable rates and bills.

This survey was very similar to one conducted by a UK-based electricity services supplier who had the same objective, i.e. to identify problem areas and customer needs. The measures were:

- supply reliability;
- price;
- speedy response to telephone calls;
- speedy response to correspondence;
- reliability in keeping appointments;
- making appointments designed to meet customer needs;
- approachability/helpfulness of staff (particularly in shops);
- fair treatment of customers; and
- accuracy/understanding of accounts.

Benchmarking – the regulator's perspective

The role of the regulators is to make sure that customers and consumers receive value for money all the time. They are empowered to monitor levels of service and also to set standards which cover aspects such as reliability of product and services supplied, the availability of what is supplied and the services surrounding the customer and company interface, i.e. those services which deal with customer complaints, customer queries and/or customer involvement.

Regulators focus on output rather than input. As such, they focus on the services and quality of products supplied, rather than the processes which deliver those and the way they are managed by organizations. They also ensure that customers are involved and that suppliers are addressing their needs efficiently and involving them in the various decisions. The regulators are not specifically interested in individual companies, but cover the industry as a whole. One of the key areas where regulators become involved is the availability of information and its transparency. The information obtained enables the regulator to conduct benchmarking exercises to identify competitive parameters and to establish standards, so the key activities that they will focus on include:

- the measurement of service delivered to the customer;
- publishing the results; and
- rewarding good performance.

The regulators, therefore, are not interested in what is measurable and what companies publish in terms of the performance measures. They are more interested in measurement from the point of view of the customer.

This transparency of information and its availability is perceived to be very important. Indeed, it is considered that the measurement of performance in itself is not enough. The results have to be made available to the individual companies concerned themselves, but also to customers.

1. The customers are entitled to know how the quality of the service they receive from particular companies compares with that provided by others, and thus enables them to arrive at decisions of future involvement with their suppliers.
2. The individual companies themselves need to know where they are in relation to other providers for them to work hard at closing the gap. It is an opportunity for them to identify best practice and incorporate it in their internal operations so that their standards match those of best in class.

This is perhaps the direct way in which regulators can apply the art of benchmarking and can influence individual organizations to close gaps and achieve high standards of performance. Florida Power and Light for instance have found that by independently and voluntarily buying in the quality ethos and by achieving high standards and setting tighter targets for themselves, they are in a position to meet minimum standard requirements set by the regulators; but they can also influence the regulators' decisions in setting standards because they have become a model company that other suppliers of utilities are aspiring to follow.

Benchmarking can also be applied by regulators to look at various standards from different industries (Saunders, 1993). Figure 12.2 illustrates results from a number of customer surveys conducted by different utilities. By compiling the surveys, an interesting benchmarking exercise could take place between the four different utility suppliers involved.

As Fig. 12.2 indicates, based on the question of whether customers are satisfied or dissatisfied with the overall service provided, British Gas come on top followed by a local electricity company, British Telecom, and lastly a local water company.

In conclusion, therefore, the art of benchmarking is applicable to utilities even though they are a very special industry.

12.2 Benchmarking in financial services

The financial services sector is undergoing substantial changes at the present time, not only in Europe but in the US as well. The recent rises in unemployment have lead to increases in house possessions and arrears, thus forcing mortgage lenders to devote substantial resources to deal with this problem. In addition, because of unemployment and worries about job security, building societies are not attracting new monies from personal investors. Other factors that are leading to great changes in financial services include:

- increasing saturation of the market, making growth much more difficult;
- growth now will not come from untapped markets because of the above, but rather from key competitors;

- because of the above, interest in quality and benchmarking has increased;
- customers are becoming much more sophisticated and more demanding;
- too much focus on product quality (most products are easy to copy, therefore achieving product quality excellence does not sustain competitive advantage); and
- customers find it difficult to distinguish between various products on offer and are unable to influence design or specification aspects.

In view of the above, benchmarking and the use of TQM principles have become increasingly important recently. The following section illustrates

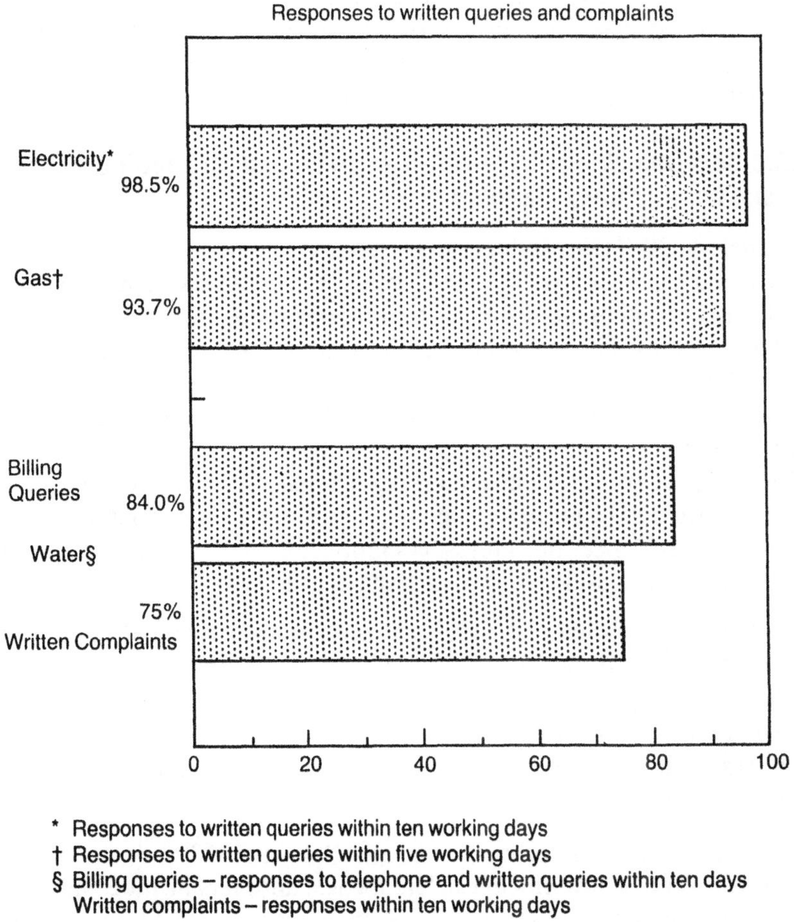

Figure 12.2 Results from customer surveys in utilities.

how benchmarking can be made applicable to financial services in a variety of ways. The examples provided refer specifically to the UK situation.

12.2.1 UBS Phillips and Drew league table

The largest 20 building societies in the UK were benchmarked using 12 different performance indicators including:

- four profitability ratios;
- two capital strength ratios;
- two market share ratios;
- two management expense ratios;
- one interest margin ratio; and
- percentage growth in assets.

The performance indicators were calculated from the analysis of the societies' most recent annual reports. Table 12.1 illustrates the results. This exercise is useful in helping individual building societies determine their strengths and weaknesses and could trigger useful benchmarking exercises to analyse the processes involved.

12.2.2 Benchmarking of TQM factors in largest 20 building societies and one bank (Abbey National)

A model of 30 TQM factors was used for this analysis. The analysis was carried out in May/June 1992, based on published annual accounts and financial statements from all the organizations involved, in relation to the previous full financial year. From these reports, information analysed included major strategic issues (internal and external), the focus on specific changes within the industry and details of senior management appointments.

The purpose of this benchmarking exercise was to classify the 20 building societies in terms of high, moderate and low adopters through prima-facie evidence of interest/commitment to TQM principles and techniques. The classification was arrived at as follows:

- **High adopters** Defined as significant prima-facie evidence resulting in a score of 15 factors or more;
- **Moderate adopters** Defined as some prima-facie evidence resulting in a score of ten factors or more;
- **Others** Defined as low or no prima-facie evidence resulting in a score of less than ten.

The information gathered showed that less than 50% of the societies have introduced TQM principles. The highest adopters include National & Provincial and Birmingham Midshires, who each scored 17, and Alliance & Leicester (Giro Bank specific) who scored 16. Moderate adopters

Table 12.1 UBS Phillips and Drew league table of building societies

	Position in 1992	Average rank	Position in 1991	Change over the year
Northern Rock	1	5.50	3	+2
Cheltenham & Gloucester	2	5.75	1	−1
Yorkshire	3	5.83	4	+1
Halifax	4	6.25	5	+1
Coventry	5	7.50	11	+6
Leeds Permanent	6	7.92	2	−4
North of England	7	8.17	6	−1
Derbyshire	8	8.33	7	−1
National & Provincial	9	9.42	10	+1
Bradford & Bingley	10	9.75	8	−2
Britannia	11=	11.50	12	+1
Woolwich	11=	11.50	15	+4
Portman	13	12.67	18	+5
Leeds & Holbeck	14	13.25	17	+3
Nationwide	15	13.75	14	−1
Alliance & Leicester	16=	13.92	13	−3
Bristol & West	16=	13.92	9	−7
Birmingham Midshires	18	14.00	20	+2
Chelsea	19	15.00	16	−3
Skipton	20	16.08	19	−1

Note: 1991 performance rankings have been recalculated so as to take into account the inclusion of North of England, and exclusion of Town & Country building societies.

include Britannia and Abbey National who scored 10, and Leeds with 12. The Halifax and Woolwich were 'traditionals'.

12.2.3 Benchmarking at the Prudential Assurance Company

The Prudential has come to realize that, because there is low differentiation between the various operators in financial services as far as products and services are concerned, brand performance is key. Branding reassurances were found to be potentially very powerful in influencing consumer choice. As argued by a Prudential Manager (Brown, 1992):

> People are in unfamiliar territory, uncertain about what they're buying, and anxious to make the right choices. . . . Yet, if you think of buying a personal pension, for example, you cannot see or touch or hear or smell it. One cannot weigh it or X-ray it or photograph it.

> The first question a consumer asks about his or her pension is 'How much will I get?' and we cannot answer that question. And yet apart from a mortgage, a personal pension is probably the largest expenditure a person will make in his or her whole life.

Understanding consumer behaviour, the brand attributes and what the consumers value most is therefore essential.

The Prudential developed a benchmarking framework (Brown, 1992) which works in two stages.

1. First they established performance of their brands against competitors using customer feedback and views on dealing with various competitors. The overall brand status is measured as a ratio of the number of customers 'happy to deal with the Prudential' over the total number of customers. Taking this ratio over time, and benchmarking the Prudential against other financial companies, gives feedback on strengths and weaknesses of brands from a consumer perspective.
2. As Fig. 12.3 illustrates, once customers start buying the brand, it is very important to retain them.

The Prudential developed a customer satisfaction grid to determine the most important elements of the service quality offered (Brown, 1992). This is then supplemented by a measurement tool based on a survey of Prudential's customers to identify:

- the value of each element of the service to customers (prioritization); and
- how Prudential is delivering each element of the service (rating).

The customer satisfaction grid can then be revisited to classify all the service elements from the customers' perspective. The measurement exercise is then used on customers of main competitors for the benchmarking exercise. The data obtained enables Prudential to benchmark its performance against all competitors, and once the various weaknesses are identified, it can then develop a strategy for carrying out the necessary improvements.

12.2.4 Corporate reporting and TQM – a benchmarking example

This study was conducted by one of the authors, looking at TQM visibility in external reporting by an analysis of corporate reports during the 1991 financial year of the following six building societies in the UK:

- Birmingham Midshires
- Leeds Permanent
- Alliance & Leicester
- Britannia

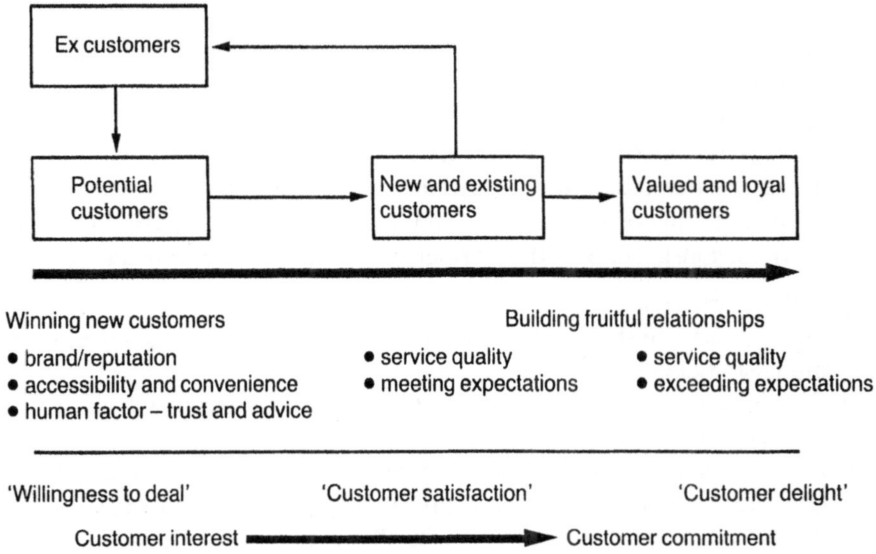

Figure 12.3 A customer relationships model at the Prudential.

- Yorkshire
- National & Provincial

Benchmarking the gross capital ratio and the increase/decrease in profits was not found to be useful. As a result of this, a framework for conducting the benchmarking exercise was developed, covering the following critical factors:

- mission visibility;
- quality policy/strategy;
- focus on customer;
- customer satisfaction commitment;
- people/employee commitment;
- process management;
- non-financial performance measurement; and
- interesting appointments.

A scale of 0–5 was used to rate each society, based on information visibility, explicit statements and evidence/examples of applications. Table 12.2 illustrates the overall outcomes of this benchmarking exercise. The National & Provincial and Birmingham Midshires societies were found to be the most committed to TQM implementation. Examples of 'best practice' found during the study are given below.

Mission visibility

- People in every part of our business now better understand that the customer comes first and that loyalty must be won through service excellence and regular communication. Birmingham Midshires' 'first choice' philosophy focuses all our people on delivering quality of service. We are committed to the continual monitoring and improvement of our performance in this area.
- To satisfy our customers' requirements for homemaking, security and protection, managing money, and providing financial products and services, consistent with the prosperity of National & Provincial.

Quality policy/strategy

- Our aim is to be the natural first choice provider of financial services for our customers.
- Our objective is to create and sustain quality standards which attract and retain customers.
- A commitment to quality is deeply ingrained in our policies and culture.
- Our aim is to provide quality products and services throughout the diversified range of businesses and to all our customer groups.
- Our aim is to continuously offer and provide superior services which meet the evolving needs of our customers and are of best value.
- Our success in realizing our direction is measured both by the financial strength of the organization – our quantitative goals – and the strength of our customer relationships – one of our qualitative goals.

Focus on customer

- Define and communicate best practices so that they are successfully implemented, incorporating the views of the customer.
- Management layers decreased, district teams increases in order to understand customer needs and to demonstrate leadership in creating the right environment to provide appropriate products and services.
- Customer first equals product, pricing and delivery terms.
- Increasingly branches are being remodelled so that customers can be served better.
- We are directing our efforts to assess and understand the real needs of our customers, and ensuring high levels of service and quality in all our operations.
- Our strategy is to understand our customers' requirements as they develop through their life in the context of homemaking, security and protection, managing money and providing for the future.

Commitment to customer satisfaction

- Priority is to maintain and improve high quality of service.
- The ultimate success is the quality of customer service. The corporate aim is strongly supported by customer first, a philosophy which places the needs and expectations of customers above all other considerations.

Table 12.2 Benchmarking in financial services: commitment to quality

	Birmingham Midshires	Leeds Permanent	Alliance & Leicester	Britannia	Yorkshire	National & Provincial
Mission visibility	5	0	0	0	0	5
Quality policy/strategy	5	2	2	0	0	5
Focus on customer	5	4	2	4	1	5
Customer satisfaction commitment	5	3	2	3	1	5
People/employee commitment	4	5	3	2	4	5
Process management	4	2	3	2	2	5
Non-financial performance measurement	4	0	0	0	0	3
Interesting appointments	5	0	0	0	2	5
Total quality score	37	16	12	11	10	38

- Our aim is to develop the relationship with our customers in order to understand and therefore anticipate changes to their requirements.
- We moved from a product orientation to focusing on our customers' requirements.
- As we move towards the achievement of our goals we will demonstrate clearly how highly we value the quality of our customer relationships.

Process management

- Newly installed technology is to enable staff to devote more time to customers and increase productivity and efficiency.
- During the year the society refurbished 63 branch offices and continued evolving branch designs which will best serve future customer needs in a cost effective manner.
- The total quality management initiative has progressed during the year and, in 1991, particular emphasis has been placed on the quality of administrative procedures. All processes are under scrutiny to ensure they are made as efficient and effective as possible for the benefit of both customers and staff.
- As our customers' requirements are constantly changing, our organizational design needs to be creative, responsive and adaptable. We have therefore created an organization based on a process approach designed to make the most of team work, which ensures that everything we do is in support of our mission.
- In 1992, we will define and document the processes through which we implement activities and thereby improve the quality of our operations.

People commitment

- To succeed, we have to enable our staff to search for continuous improvement and to think for the future, both of which will benefit our customers.
- Our quality award scheme recognizes outstanding customer service, and encourages staff to take responsibility for finding solutions to customer problems.
- Passing authority at branch level (relaxing sensible and well proven controls through considerable training of personnel).
- The modern branch manager must be a catalyst, coach and team leader.
- Investment in people is a crucial part of the society's development.
- Our aim is to strengthen the society through teamwork, continuous challenge and improvement thereby sustaining the quality which ensures competitiveness.
- Without the right quality of players in our teams we cannot succeed. We see the relationship between an individual and the organization as a partnership with responsibilites on both sides.

Non-financial performance measurement

- Quality of service has improved – 95.5% of customers are highly satisfied or satisfied.
- We measure the effectiveness of everything we do to ensure that we learn and continually improve.
- Our success in realizing our direction is measured both by the financial strength of the organization – our quantitative goals – and the strength of our customer relationships – one of our qualitative goals.

Interesting appointments

- Head of customer services
- Head of credit risk and quality assurance
- General manager (business development)
- Business consultant
- Director of business and organization development
- Director of customer requirements

12.3 Benchmarking in innovation

Interest in innovation is growing at a staggering pace. In the 1970s and 1980s, innovation tended to be product based and competitive advantage tended to go hand in hand with technical know-how. Benchmarking in innovation at that time was difficult to carry out because technologies determined the extent of competitive gaps between various competitors and therefore they had to be protected. In the 1990s, however, the technological gap between various competitors has gradually been closing. New technology, its ease of adoption and relative low cost meant that entry into the market became less difficult; copying of complex products became an easy task. Organizations therefore started to look at new means of establishing superiority and sustaining leading positions in the marketplace. Focusing on customers and delivering customer wants became the major task. Speed, quality, consistency and reliability of delivery, and cost became the key elements of determining competitive standards. Organizations therefore became less nervous about sharing information on product technologies and are finding information on the process of developing new products with speed, quality, very sensitive to share. Many major benchmarking studies on **product** innovation took place and are still taking place. Amongst the biggest are the following:

- the SAPPHO project;
- project NEWPROD;
- Stanford innovation project;
- Souder study; and
- the AMT project.

The SAPPHO project

This study was carried out in the 1970s (Rothwell *et al.*, 1974) by comparing 43 pairs of innovations representing the instrumentation and chemical industries. The project identified five main factors thought to lead to the successful diffusion of innovations:

1. a very good understanding of user needs;
2. marketing treated with a lot of attention;
3. development work performed with care and attention;
4. positive usage of existing technology and outside service; and
5. quality of responsible people making decisions.

Broadly speaking, the SAPPHO project identified characteristics representing three main areas:

- organizational factors;
- managerial factors; and
- product development factors.

Project NEWPROD

This study was also carried out in the late 1970s (Cooper, 1979, 1980) using a sample of 195 industrial manufacturers involved in new product development. The study identified three factors which strongly facilitate the diffusion of new products:

1. product superiority and excellence in the eyes of the customer;
2. proficient, systematic and strong knowledge of the market;
3. proficient combination of technical and product synergy.

The identified factors refer to two main areas:

- product factors; and
- market factors.

Stanford innovation project

This study was conducted in 1982 (Maidique and Zirger, 1984) to explain the differences between the SAPPHO project and project NEWPROD. Its main findings relate to eight key areas of success for new product development in a highly technological environment:

1. depth of market knowledge achieved through successful interaction with customers;
2. the planning of the new product process, in particular the R&D stage;
3. the co-ordination of the new product process with special emphasis on R&D;
4. the special attention given to marketing and sales effort;
5. management commitment and support from the onset, i.e. from the development to the launch stage;
6. the contribution margin of the product;

7. early entry in the marketplace; and
8. the close match between the new product technologies, the market considered and the existing strengths of the supplying company.

Broadly speaking the factors identified by the Stanford project as conducive towards success in the diffusion process can be condensed into five main areas:

- market characteristics;
- management characteristics;
- organizational characteristics;
- new product development characteristics; and
- product characteristics.

The Souder study
This is a more recent study which looked at success criteria in new product innovations (Souder, 1987). The factors identified which lead to successful diffusion include:

1. the high level of understanding of user needs and the technical expertise in helping solve problems;
2. the strong relationship between the level of expertise available and the technology being marketed; and
3. the excellence in the quality and level of support given.

These factors have been related to two main areas:

- managerial characteristics; and
- organizational characteristics.

Table 12.3 illustrates lists of criteria which distinguish innovative organizations from non-innovative ones, as found from the Souder study. Table 12.4, on the other hand, shows seven critical factors of innovation and how organizations deal with each one (Souder, 1987).

The AMT project
This study, conducted by one of the authors (Zairi, 1990), looked at suppliers' strengths and weaknesses in the diffusion of advanced manufacturing technology (AMT). Seven different types of technological systems were examined in ten different supplier bases. The following factors were found to lead to successful diffusion of AMT:

1. price of AMT and the vast range of products which can cater for various customer needs;
2. existing features, the level of technology incorporated and the flexibility of AMT equipment;
3. the ability of products to perform to required standards;
4. the quality of support given to customers;
5. the size of organizations and vast resources available;

Table 12.3 Classical versus innovative organizing principles

Classical	Innovative
Jobs are narrowly defined and subdivided into rigid, small units	There is constant adjustment of tasks through the interactions of the organization members
A narrow definition of authority is attached to the individual's job	A sense of responsibility replaces authority and there is commitment to the organization that goes beyond the individual's functional role
There are many hierarchical levels and a strict hierarchy of control and authority; the bosses order things to be done	There is a low degree of hierarchy of control and authority; things get done through a mutuality of agreement and a community of interest
Communication is mainly vertical between superiors and subordinates; these communicaions consist of instructions issued by superiors	Communication runs in all directions between people of different ranks and it resembles consultation rather than command
There are many rules and policies, and loyalty and obedience to superiors are required	Emotional commitment to the achievement of tasks and the expansion of the firm is highly valued
Economic efficiency is the goal	Newness and creativity are sought and growth is the goal
Top-down authority-centred management is the mode	Horizontal, expertise-centred influence is the mode; teams, task forces, and project management methods are used

Source: Souder (1987)

6. the ability to work closely with other suppliers;
7. project management capability and quality of personnel employed; and
8. quality of communication systems.

Broadly speaking, these factors can be grouped into three main areas:

- product characteristics;
- organizational characteristics; and
- managerial characteristics.

Table 12.4 Seven characteristics of innovations and the corresponding qualities of innovative organizations

Seven characteristics of innovations that raise hurdles	Corresponding qualities of innovative organizations required to cope with these hurdles
Disruptiveness; they create high sociobehavioural costs	Willingness to accept change, altered behaviours, and disruptions
Involve relatively high research, development, and commercialization costs	Long-term commitment to technology
Require time for idea germination, gestation, and maturation	Patience in permitting ideas to gestate and decisiveness in allocating resources to those ideas having the greatest commercial prospects
Carry high risk of uncertainty	Willingness to confront uncertainties and accept balanced risks
Timeliness	Alertness in sensing environmental threats and opportunities and promptness in responding to them
Involve combinations of various technologies that may exist inside and outside the firm	Openness of internal, cross-departmental communications; diversity of internal talents and cultures; existence of many external contacts and information sources
Involve the talents of many individuals who collectively possess interdisciplinary know-how	A climate that fosters the natural confrontation and resolution of interdepartmental rivalries and conflicts and the development of reciprocal role-persons

Source: Souder (1987)

Benchmarking in product innovation has led to the establishment of very useful comparisons and the identification of critical factors for product innovation, identified commonly by all the studies. The following are outcomes from the various comparisons:

1. all the studies, with the exception of NEWPROD, have emphasized the importance of organizational and managerial characteristics;

2. most studies have highlighted the importance of communication processes with customers;
3. with the exception of SAPPHO, all the studies have stressed the importance of matching closely the level of technology to suppliers' strengths;
4. the NEWPROD study and the AMT project highighted the importance of product criteria; and
5. with the exception of SAPPHO and the Stanford product, studies did not place major emphasis on R&D and new product development.

12.3.1 Benchmarking process innovation – the McKinsey model

Innovation in the 1990s refers to the business delivery **process** in its totality and not just specifically the products and services which are developed and then launched into the marketplace. In this situation, benchmarking becomes a much more difficult task, since any instrument used for benchmarking exercises has to be an all-encompassing one. One of the most powerful instruments developed is based on the McKinsey model or the 7S model, so called because the premise behind the McKinsey model is that for organizations to function effectively they have to rely on the interdependence of seven variables all beginning with 's'.

- **Strategy:** the plan leading to the allocation of resources.
- **Share values:** the goals shared by all employees.
- **Style:** the management style of the organization.
- **Structure:** the organizational map/chart.
- **Skills:** the strengths and capabilities of all employees.
- **Staff:** the people employed.
- **Systems:** procedures, guidelines and control mechanisms.

The seven variables are classified as:

1. **hardware variables** (strategy, structure); and
2. **software variables** (style, systems, staff, skills and shared values).

The major contribution of the framework is the attention it draws to the less tangible, less visible aspects of organizational systems. As explained by Peters and Waterman (1982):

> In retrospect, what our framework has really done is to remind the world of professional managers that 'soft' is hard. It has enabled us to say, in effect, all that . . . you have been dismissing for so long as intractable, irrational, intuitive, informal organization can be managed. Clearly, it has as much or more to do with the way things work (or don't) around your companies as the formal structures and strategies do.

This was further explained by Waterman, Peters and Philips (1980):

> Our assertion is that productive organization change is not simply a matter of structure, although structure is important. It is not so simple as the interaction between strategy and structure, although strategy is critical too. Our claim is that effective organizational change is really the relationship between structure, strategy, systems, style, skills, staff, and something we call superordinate goals.

The McKinsey model (Fig. 12.4) is sometimes referred to as the happy atom. It reflects the following characteristics:

1. **multiplicity of factors** – all influence how organizations behave;
2. **interconnectedness of variables** – progress can only be achieved by giving attention to all areas;
3. **all seven variables act as a driving force** – at particular points in time, one or more of the seven 's's will emerge as the most critical variable(s).

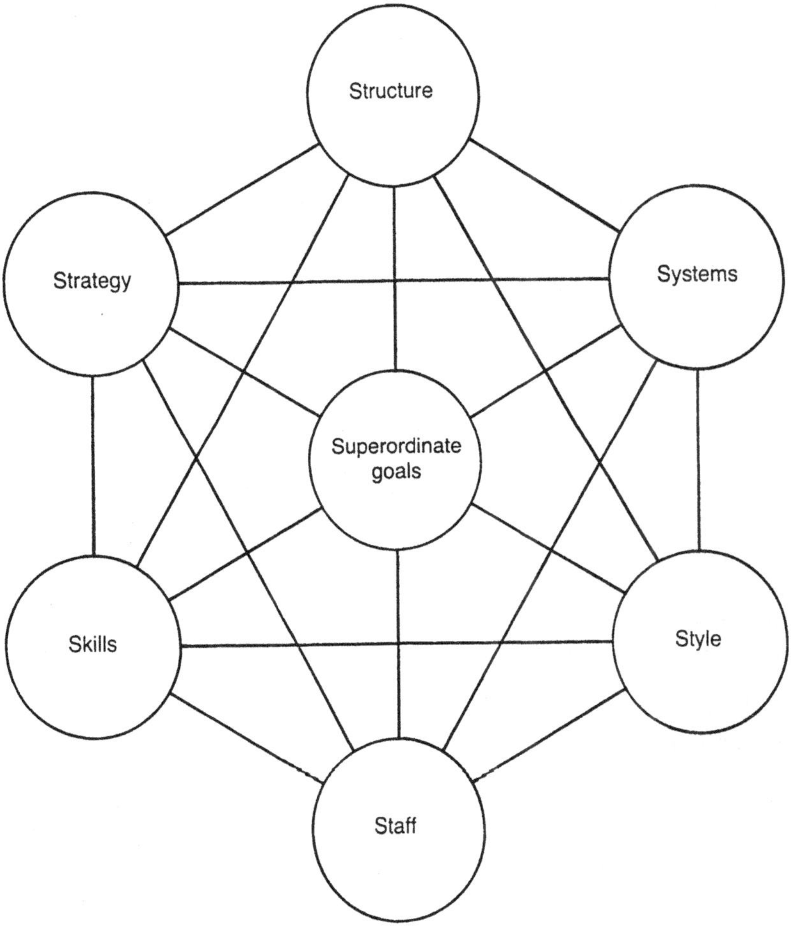

Figure 12.4 The McKinsey model.

Origins of the McKinsey model

Many researchers have commented on the inadequacies of structure as a way to represent a dynamic organization. As a result of this the McKinsey group developed an alternative framework. The 7S model was tried in teaching and workshop situations and also piloted on a small number of companies. The model was found to work as a diagnostic tool for assessing organizations' competitive performance and as an aid to managers in formulating action plans and improvement strategies.

Applicability of the McKinsey model to the area of innovation

The 7S framework has been used very successfully in the area of innovation research. Work on innovation carried out by Pascale and Athos (1982) and later by Johne and Snelson (1988) was based on the 7S model. Indeed innovation processes can be described by hard and soft organizational variables as follows.

- **Strategy:** Is there a product development strategy which defines types of projects selected and the resources required?
- **Style:** Is there top management commitment and how much support is there from the top for new product development?
- **Shared values:** How much belief, enthusiasm and commitment is there for innovating activity?
- **Structure:** What lines of authority and responsibility are used for innovating activity?
- **Skills:** What specialist knowledge, tools and techniques are used for innovating activity?
- **Staff:** People involvement, empowerment, teamwork and degree of participation in decision making in relation to product development?
- **Systems:** What procedures, guidelines and control mechanisms are used for managing innovation activity?

Using the 7S model to benchmark innovation processes – an example

Using similar questions to those above, Johne and Snelson (1988) examined innovation activity in 20 UK and 20 US companies. The companies represented mechanical and electrical engineering sectors, chemical and food industries.

The purpose of the study was to establish:

- which firms are competing successfully and relying on an active product development strategy; and
- how do their innovation practices compare with those of less successful companies who do not use innovation strategies effectively?

The results indicated that the seven areas suggested by the McKinsey model are a combined responsibility of top and middle management.

- **Top management:**
  - determine the product development strategy;
  - create a climate of shared belief for introducing change;
  - give active support to the various teams involved in new product development; and
  - develop a suitable structure for managing innovation activities.
- **Middle management:**
  - choose the relevant skills required;
  - select people with the most appropriate knowledge for the tasks involved; and
  - ensure that the most appropriate control mechanisms and right procedures are being used.

A summary of the key findings of the Johne and Snelson study shows that in new product development, leading innovators have top management much more involved in the process:

- they set broad objectives for organic growth (strategy);
- they foster understanding of the need for innovative products (shared values);
- they are intimately involved, often on a day-to-day basis (style); and
- they use new methods and structures, e.g. business teams (structure).

Less successful innovators have top management more distracted from the innovation process. Their top management:

- prefers to catch up by acquisition rather than by organic growth (strategy);
- accepts that the existing organization will feel threatened by any radical product developments which might undermine existing powerbases (shared values);
- only involves itself directly in acquistions, leaving new ventures and their considerable risks to individuals within the organization (style); and
- takes new product develop completely outside the existing formal structure (structure).

In addition, less successful innovators tend to rely more on outside consultants and less on new technology than their successful counterparts who utilize in-house skills, encourage staff creativity and take full advantage of all that new technology has to offer.

The Zairi benchmarking study of innovation processes

This study looked at innovation processes of twelve of the leading performers in European and world markets. An adapation of the 7S model was used to develop a comprehensive benchmarking instrument.

Using the McKinsey model criteria (strategy, structure, systems, shared values, staff, skills, style), a variety of good practices were identified covering many aspects of managing innovating activity:

1. having top management playing a strategic role in directing, facilitating the allocation of resources and being active in reviewing and planning innovation activity;
2. having innovation activity as an integral part of corporate strategy;
3. top management commitment in creating a positive climate for innovation and actively supporting all the processes of innovation;
4. having innovation as a voluntary activity, and a firm belief that innovation is vital to an organization's ability to remain competitive;
5. having effective communication processes from the corporate level downwards, with clear objectives and a thorough understanding of the organizational goal – communication also includes sharing information on results and action plans;
6. having a participative style of management, with a distributed approach to decision making and full support from top management;
7. project-based structure of managing innovation with multi-disciplinary teams, formal and informal reporting mechanisms and measurement;
8. project management driven by a thorough understanding of customer requirements, process capability and the organizational goal;
9. organizations using all skills at their disposal effectively as and when required;
10. project management, not individually led, but driven by the creative contributions of all functions within the organization;
11. use of modern tools and techniques in managing innovation activity;
12. systems which are vital to effective management of innovation activity and are important for setting up the goals, managing the performance at individual project level and also at business level; and
13. systems in place to track down killer variables and to enable project leaders/teams/senior managers decide on whether to terminate or proceed with projects.

12.4 Benchmarking in human resources

Most leading companies have recognized that if proper attention is devoted to the human resource issues, this is often followed by high productivity and better competitiveness. The importance of people is often highlighted in mission statements, quality policies and strategies.

In the following leading companies, the human resource issue is made very visible and is at the heart of any drive for carrying out improvement efforts.

1. **Florida Power and Light,** a winner of the Deming Prize, bases its quality improvement strategy on four key principles:

- customer satisfaction;
- PDCA (plan–do–check–act);
- management by fact – first collect objective data, second, use the information to solve problems and improve work quality; and
- respect for people.

2. **Cadillac,** a winner of the 1990 Baldrige award, expresses its commitment to people in the mission statement:

 The mission of the Cadillac Motor Car Company is to engineer, produce and market the world's finest automobile known for uncompromised levels of distinctiveness, comfort, convenience and refined performance. Through its people, who are its strength, Cadillac will continuously improve the quality of its products and services to meet or exceed customer expectations and succeed as a profitable business.

3. **Marlow Industries, Inc.,** a winner of the 1991 Baldrige award, in its corporate vision, states its commitment to ensuring total employee satisfaction:

 Our corporate vision defines the fundamental purpose of our company:
 - provide thermoelectric products, and services that will meet or exceed the customers' requirements, without exception;
 - provide an environment that encourages our employees to achieve a high degree of productivity through job satisfaction;
 - provide growth in revenues and net worth for the shareholders at a controlled rate demonstrating financial stability.

Many more statements could be referred to where there is visible acknowledgement and appreciation that TQM efforts and improvement programmes can only be carried out through the right channelling of people's efforts. **Kodak Limited,** for example, expanding on a statement which recognizes people as their most important asset, have the following policy to show true commitment:

> To recognize that our people are our most important asset, we will have policies and practices which will:
>
> - provide standards of health and safety and environmental performance such that the company is classified amongst the best when compared with appropriate sectors of the industry;
> - afford equity of treatment to all employees;
> - provide effective channels of communication with employees;
> - ensure provision of overall employment package which is equal to, or better than, that offered by comparable businesses in the community;

- aim to provide, in partnering with its employees, stability of employment; and
- assist in improving the overall performance of its people.

BP Chemicals recognize people as one of their six pillars of strength. 'The high calibre of our staff is one of our major strengths, and it is vital that we continue to attract, retain, motivate and develop the best people.'

The recognition of people and their valuable contribution indicates that companies have shifted from relying too much on technological sophistication, rigid control systems, and total reliance on the method with people considered as an after thought in the process, a bolt on. The quality movement has certainly helped in highlighting the importance of people. In addition, the emphasis being based on working smarter rather than harder means that people are all the more important because of their flexibility and their ability to assume various different roles and absorb new knowledge required for dealing with new change.

12.4.1 The importance of the human resource element

To understand the importance of the human resource element, it is necessary to refer to the various frameworks which are used for assessing the overall effectiveness and competitiveness of organizations. Such frameworks include for example the Deming Prize, the Malcolm Baldrige National Quality Award (MBNQA), the European Quality Award (EQA) amongst others.

Benchmarking human resource practices using the Deming Prize model

The Deming Prize was introduced in 1951 by the Japanese Union of Scientists and Engineers (JUSE) in recognition of the efforts of Dr W. Deming and his efforts in putting the Japanese quality standards on track. The prize is very prestigious and considered to be important for most organizations wishing to improve their performance standards in all aspects covering products, processes, services.

The Deming Prize criteria used for assessment include ten different areas covering the following:

1. Policy and objectives
2. Organization and its operation
   - 2.5 utilization of staff
   - 2.6 utilization of QC circle activities
3. Education and its extension
   - 3.1 Education plan and actual accomplishment
   - 3.2 Consciousness about quality control, understanding of QC
   - 3.3 Education concerning statistical concepts and methods, and degree of permeation
   - 3.4 Ability to understand effects

3.5 Education for subcontractors and outside organizations
3.6 QC circle activities
3.7 Suggestion system and its implementation
4. Assembling and disseminating information, and its utilization
5. Analysis
6. Standardization
7. Control (*kanri*)
8. Quality assurance
9. Effects
10. Future plans

Florida Power and Light was the first non-Japanese winner of the Deming Prize, in 1989. Human resource practices which led them to win included the use of quality improvement (QI) teams. The work of quality improvement teams is shared in various forms. QI teams meet for about an hour every week. They report to their TQ facilitator and to the quality council. Participation in the teams is on a voluntary basis.

In addition there are multidisciplinary task teams chosen by management to tackle bigger problems.

All employees are trained in the use of seven basic tools of problem solving, i.e. check sheets, histograms, pareto diagrams, cause and effect, graphs, scatter diagrams and control charts. The company insists on the tools being used for team problem-solving activity. In addition to these seven tools, the company uses a problem-solving process as follows:

- reason for improvement (what seems to be the problem?);
- current situation (where are we now?);
- analysis (what are the root causes of the problem?);
- countermeasures (what should we do to improve the situation?);
- results (what happened as a result of our actions?);
- standardization (how do we hold on to the improvement?); and
- future plans (what is the next item we need to work on to achieve further improvement in this area?).

As well as the teams, Florida Power and Light uses 'quality in daily work' (QIDW) as a way of managing. It focuses on doing things right and doing the right things using the customer–supplier links principle. It establishes the principle of internal customers: who are they? what are their requirements? what measures are used for managing deliverables to those customers? what is the most important thing which is out of control, or poorly delivered to those customers?

Then it uses flow charting of processes, development of key measures, measuring and allocating consistent management of each process in order to remain capable.

The existing suggestion scheme was replaced because it is focused on individuals. 'Bright ideas' is their new scheme which generates ideas from all individual members. It is then left to the team itself to select the best, workable ideas. The company changed all its management philosophy to be based on teamwork.

Bonus, promotion and other human resource systems at Florida Power and Light are geared to recognize team quality improvement programmes (QIP). Other forms of recognition include invitations to present QI stories at various levels and various national and international forums, including invitations to families and spouses.

Benchmarking human resource management (HRM) issues using MBNQA
MBNQA is an award for American companies only, unlike the Deming prize which is offered to overseas companies as well. The purpose of this award is to promote awareness of quality as an integral part of competitive strategy, to highlight to American companies the key elements of a superior quality programme and to share information on successful outcomes from implementing effective quality strategies.

The Malcolm Baldrige framework incldes a set of values and concepts represented in seven specific categories, one of them specifically referring to the management of human resources.

1. Leadership
2. Information and analysis
3. Strategic quality planning
4. Human resource development and management
5. Management of process quality
6. Quality and operational results
7. Customer focus and satisfaction.

The human resource development and management area looks at how people are developed and involved to realize their full potential using a company's quality objectives and also examines how companies are determined to maintain a climate where there is continuous employee development, full participation and involvement towards realizing future company goals.

The human resource area has five sub-headings.

- *4.1 Human resource management:* the company's overall HRM plans and practices to support the quality improvement drive and also the spread of those practices and degree of involvement and the levels concerned.
- *4.2 Employee involvement:* the extent of involving people and the mechanisms which encourage involvement and also the levels of involvement.
- *4.3 Employee education and training:* the recognition of quality training, the method administering the training, the effectiveness in using

skills/knowledge acquired and the training received by category of employee.

- *4.4 Employee performance and recognition:* the various methods used for encouraging, giving feedback to, rewarding and recognizing employee contributions in the quality drive and in achieving the desired performance standards.
- *4.5 Employee well-being and morale:* how companies maintain and further develop a climate where employee satisfaction is ensured through their continuous development and growing involvement.

Marlow Industries, Inc., the winner of Baldrige in 1991, use the following tactics.

1. *Employee training*
   - A training action team is put in place to co-ordinate all the training activities.
   - Training covers a professional qualification system (PQS) and the Marlow employee effectiveness teams (MEETs).
   - PQS addresses job skills requirements and MEETS works through using knowledge and skills on an everyday basis.
   - Quality improvement efforts are carried out using team formats.
   - Corporate action teams (CATs), department action teams (DATs) and MEETs are allocated projects and resources and overseen by the quality council. Each CAT has a mentor, usually a member of the TQ Council.
   - Individuals are encouraged to offer suggestions and to answer the annual employee quality survey.
2. *Employee empowerment*
   - Employees are trained and encouraged to take individual action if quality of products and services is questionable.
   - Control charts are used to assist employees in making decisions about the processes they own.
   - Business is divided into market segments which are represented by multifunctional teams fully empowered to take decisions about optimizing quality of service and total customer satisfaction for their particular business market segment.
3. *Employee recognition*
   - Recognition is thought to be important at Marlow because it is key to creating and maintaining quality consciousness and motivation.
   - Company newsletter is used as a vehicle for recognition. In addition, there are other means such as hall of fame, perfect attendance, service awards, good housekeeping, clean act.
   - The company places strong emphasis on daily individual praise.

Cadillac, winner of Baldrige in 1990, use the following tactics.

- A people strategy introduced to address the specific needs of its employees and at the same time to achieve the business objectives.
- Business plan driving the people strategy, to continuously channel through people's skills and knowledge to meet business requirements.
- An integral link between people strategy and the business plan, ensures that employees are fully developed, involved in the decision-making process and kept fully informed on business priorities.
- Appropriate recognition and reward systems are linked to the people strategy in line with the business plan, so that right behaviours are exhibited to make the plan achievable.
- The structure of people strategy (Figs 12.5–6) is such that individual employees are supported by seven people strategy teams, multifunctional in nature and responsible for the research, design, recommendation, implementation of Cadillac's people processes.
- People strategy teams receive guidance on policy from an HRM operating committee.
- At the bottom level, the HRM policy committee represented by executive staff, sets the policy, direction and resource levels.
- Through simultaneous engineering, involvement is encouraged at all levels and in all the functions of the organization.

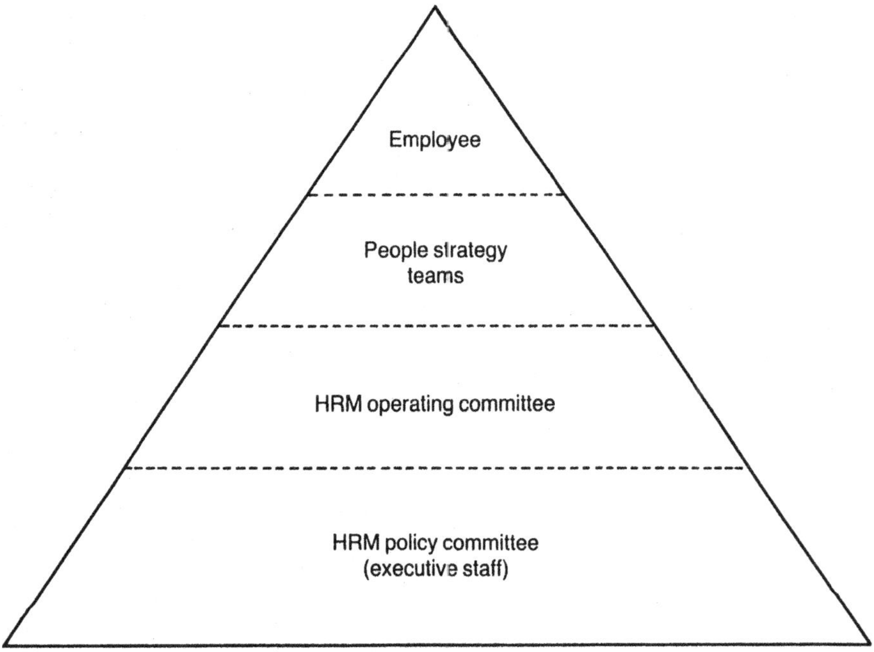

Figure 12.5 The people strategy structure at Cadillac.

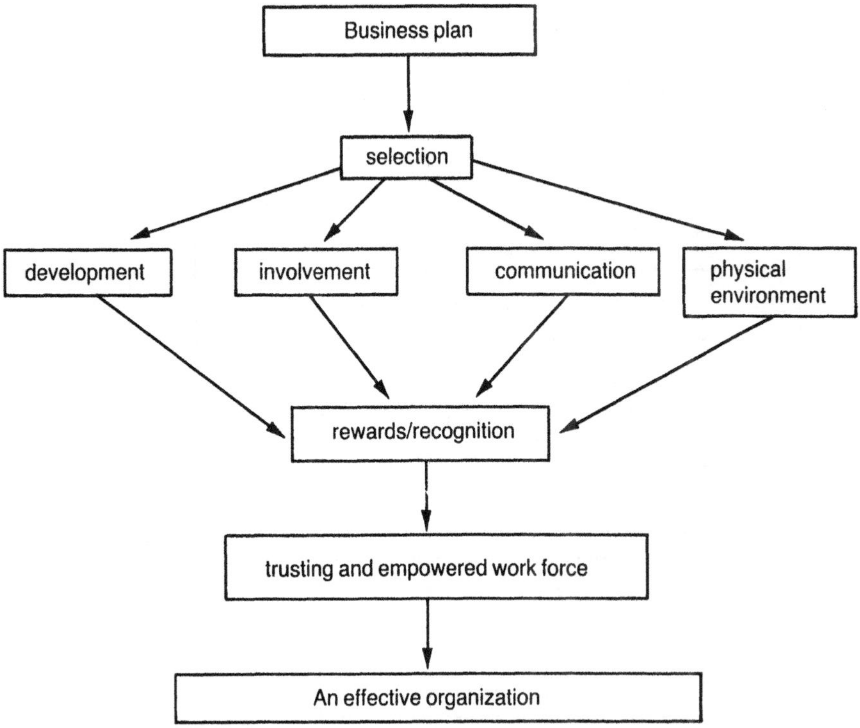

Figure 12.6 The people strategy process at Cadillac.

Through wide involvement and employee empowerment, Cadillac saw education and training levels exceed 40 hours per employee per year, injury and illness rates improved by more than 33%, employee turnover registered at one of the lowest in the industry at 0.3% per year.

IBM Rochester, winner of Baldrige in 1990, also believes that employee strategy has to be an integral part of business plans (Fig. 12.7). This strategy strives to change the culture from a product driven one to a common focus on the end customer (i.e. market driven). This is achieved through formal training and education, on the job customer contact and full participation. IBM believes that measures of effectiveness should include such things as morale, the quality of people/ideas brought in and participation effectiveness. There is a belief that human resource processes should be continuously monitored and improved in the same way as other processes.

Milliken, winner of Baldrige in 1989, apply the following principles.

- Emphasis is on finding the best people for every position and on continuing education, appraisal, development, and recognition.
- Employees are all called associates – this is because all contributions are considered valuable and there is a common objective.

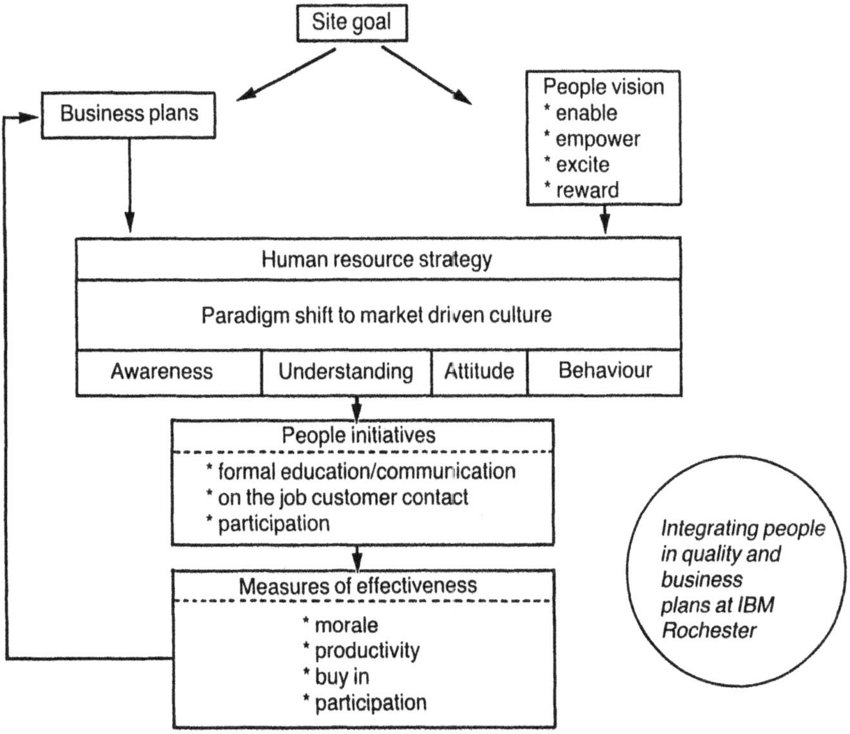

Figure 12.7 **Basic belief: respect for the individual.**

- Milliken relies on teamwork and effective communication for carrying out all activities.
- Recognition is an essential part of Milliken's philosophy, both at individual and team level.
- A participative management style ensures involvement at all levels.

Wallace Co, Inc., winner of Baldrige in 1990, develops quality strategic objectives (QSOs) every year that drive the quality process. One of the QSOs is on human resource management.

- HRM strategy includes both short-term and long-term plans; short-term plans include responding to each new suggestion within one day, increasing the quality skills inventory of all associates; long-term plans include strong career/education development for all associates.
- Associate involvement has evolved from quality circles in 1985 to fully empowered quality improvement process (QIP) team network.
- QIP teams solve problems, streamline job processes, monitor cycle time reduction, and improve processes in all divisions.
- Team participation is voluntary – QIP team activities are monitored by the QIP team coordinating board.

- There is recognition of participation and involvement: suggestions are cited regularly, citations are made in the company newsletter, and congratulatory letters sent by the CEO.
- Company looks after the well-being of its employees through employee assistance programs (EAP), tuition reimbursement plans and personal time plans for associates to use to attend school tutoring, doctor appointments and family emergencies.

Zytec, winners of Baldrige in 1991, has long- and short-term strategic objectives in terms of its HRM requirements. These are derived from its long-term strategic plan which is expressed as follows:

> Implement self managed work groups in which employees make most day-to-day decisions while management focuses on coaching and process improvement

The long-term objective is that managers will be trained to become better coaches/facilitators – the short-term one is that managers will become facilitators of self-managed work groups which are cross-functional in nature, and granted broad authority to achieve both team and individual goals.

Zytec encourages wide involvement and participation for quality improvement by providing employees with opportunities to learn and grow. Production workers designed a system called multifunctional employee (MFE) to be used by production workers. This is essentially a scheme for broadening the skills of production workers through problem solving and has a direct impact on pay. Non production workers develop personal action plans with their managers, which focus on improved performance.

Solectron, winner of Baldrige in 1991, use a continuous learning programme for all employees including the chairman of the company. HRM programmes are developed for all employees who receive training in the form of classroom contact, on the job training, mentoring programmes, communication skills, team building, etc. In addition, Solectron supports employees in furthering their education in other institutions. The training investment in Solectron is believed to have led to higher productivity and efficiency, to better bottom-line results, greater satisfaction and a drop in the levels of absenteeism. Accident rates have also fallen.

Benchmarking human resource issues using the EQA framework

The European Quality Award (EQA) was given to the first company in 1992. Essentially it was introduced to promote quality excellence in western Europe and particularly to encourage the need for self-examination/assessment on a regular basis by companies, so that areas of weakness are constantly identified and improved upon.

Unlike the Deming Prize or the MBNQA, the European Quality Award is divided into enablers and results (i.e. the hows and whats). There are nine criteria in the model, including customer satisfaction, people (employee) satisfaction, and impact on society. These are often achieved through: leadership (driving), and policy and strategy, people management, resources and processes, the end result being business results.

The human resource element under this framework is represented in terms of hows (enabler elements):

- how continuous improvement in people management is accomplished;
- how the skills and capabilities of the people are preserved and developed through recruitment, training and career progression;
- how people and teams agree targets and continuously review performance;
- how the involvement of everyone in continuous improvement is promoted and people are empowered to take appropriate action;
- how effective top-down and bottom-up communication is achieved;

and whats (measures) related to training, education, communication, effectiveness of vision, values and policy, recognition and rewards schemes, involvement, appraisal and quantitative measures such as staff turnover, absenteeism, etc.

Rank Xerox Ltd, winner of EQA in 1992, changed their culture from a traditional style of directing to a fully participative style, to gain people's commitment to carry out improvement efforts and strive for excellence. People satisfaction was established as a corporate priority, second only to customer satisfaction.

Rank Xerox determined that improvements in people management can only be carried out through improvements of the managers themselves, so a whole strategy of selecting managers, training them, assessing and developing their effectiveness was put in place. Employee satisfaction surveys (ESS) gauge trends in people satisfaction.

Goals and company objectives are shared with everyone within the organization. In addition, individual appraisals and reviews present a further opportunity for employees to be involved.

Skills are continuously developed according to the requirements of the business plan. The skills mix required is obtained from internal transfer, promotion and external hiring. The company uses skill profiles, career maps outlining long-term positions and career routes.

Training too is a continuous process consisting of short-term specific skills requirements and long-term employee development needs. Flexibility is offered to employees and dialogue with their managers determines the most effective route beneficial to both the individual concerned and the organization.

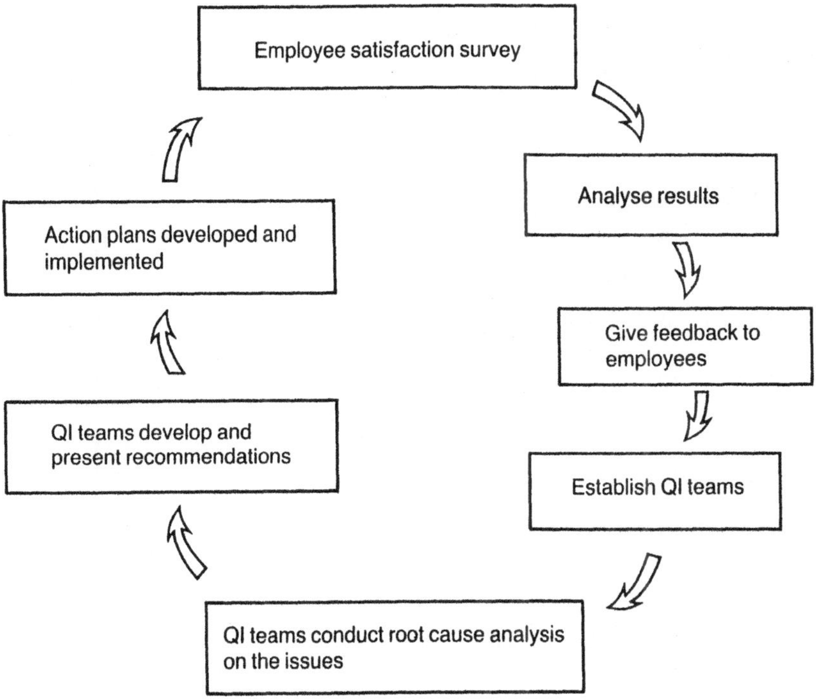

Figure 12.8 Employee satisfaction management process at Rank Xerox Ltd.

Training effectiveness is measured through course evaluation questionnaires, continuous review/annual appraisal, employee satisfaction surveys and customer satisfaction measurement surveys (CSMS).

Employees have an agreed set of annual objectives and understand the criteria against which they will be appraised. Progress against objectives is reviewed regularly and appraisal for each individual is carried out annually against the agreed objectives. As a result of the appraisal, a development action plan is worked out by both the employee concerned and his/her manager.

People involvement in quality and continuous improvement takes place through suggestion schemes operating at all levels and the use of quality improvement (QI) teams for solving problems. There are hundreds of QI teams in place, tackling a wide variety of problems at local, regional, national, and international level.

Empowerment is given to all the QI teams to provide recommendations, and to develop, implement and monitor solutions to problems. Rank Xerox introduced a programme called **capacity to act** to extend the empowerment process.

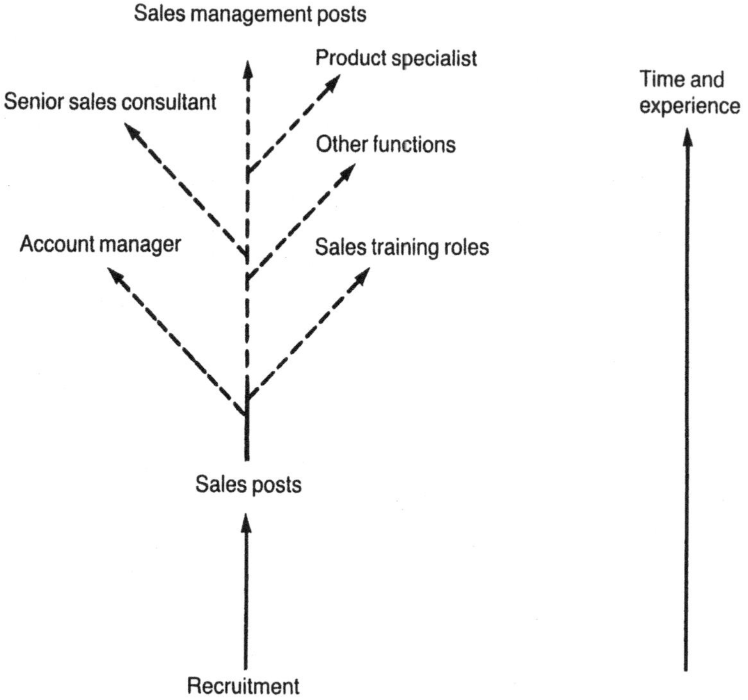

Figure 12.9 Example of career progression map at Rank Xerox Ltd.

Benchmarking human resource issues using the NASA framework
Rockwell Space Systems Division, winner of the NASA Award in 1990, use a crossfunctional approach for problem solving and conducting quality improvements.

- Communication of performance takes place through display of monthly charts. There are also product quality improvement logs at process level.
- Managers and workers review product quality performance data each day to monitor the effectiveness of process controls.
- Similar information is also used by cross-functional product quality improvement councils in each department to assess problem prevention and corrective action.
- The company has a policy of training people in multiskills so that there is flexibility in assigning tasks.
- There is a proper procedure for assigning people to tasks, based on training levels and certification of standards achieved.
- Managers are trained in a wide variety of skills and in what is referred to as the team excellence processes.

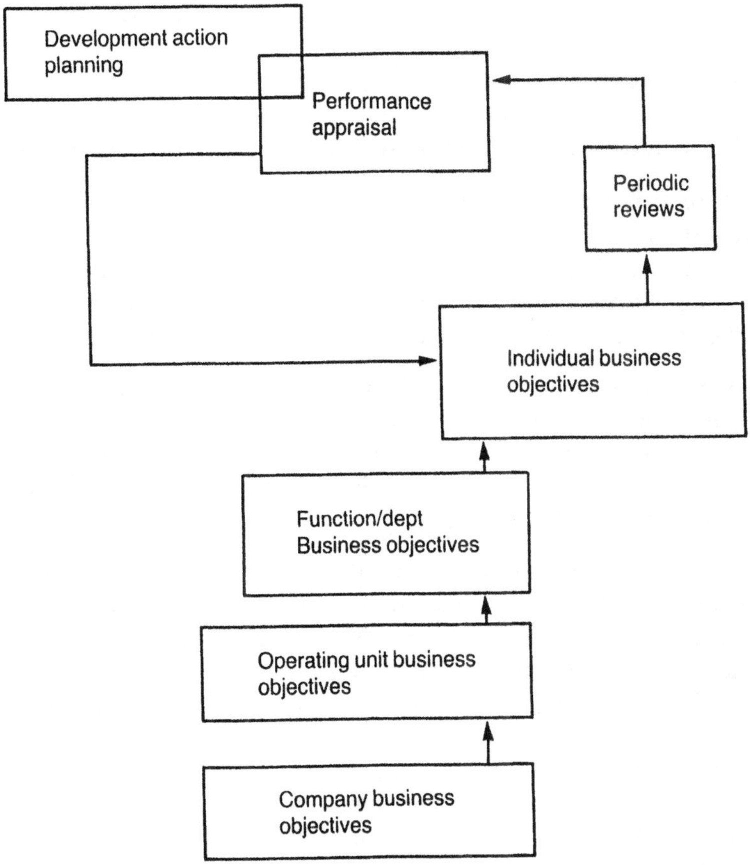

Figure 12.10 **The link between employee goal setting, appraisal and policy deployment at Rank Xerox Ltd.**

- Through a top-down and a bottom-up approach, goals are communicated, to become specific management and functional/individual performance standards.
- Team excellence initiatives are geared towards improving quality, reducing cycle time, lowering customer life cycle costs and enhancing all aspects of processes.
- Once a year, all salaried employees are expected to sit with their managers/supervisors, in order to agree a set of objectives in line with the business plan.
- Feedback and effective communication on performance takes place in various ways, through employee appraisal, monthly displays of performance and employee feedback in terms of awards and recognition programmes.
- There are three levels of team activity. In addition to the team excellence councils, there are product quality improvement councils

(cross-functional teams to prevent and conduct corrective actions) and employee improvement teams.

- A suggestion scheme programme is in operation.

12.4.2 Benchmarking the personnel function: best practice in Japan

The nature of Japanese competitiveness revolves very much around people and how their creative talents and efforts are used. This is recognized by many experts on Japanese competitiveness. Ballon (1992) writes that one of the impediments for foreign companies to operate effectively in Japan is the difficulty in hiring good people and managing them. He argues that:

> It is however, a system of human resource utilization that poses special problems for the foreign company seeking to operate in Japan . . . issues of human resource management have become the most formidable barrier to effective Japanese operations, far outweighing legal or administrative barriers.

At the heart of Japanese competitiveness is total reliance on people productivity. People are managed as a critical resource which appreciates with time and which is at the heart of strategic planning. Increased flexibility in people, their movements, their contributions both at individual and team level constitute a competitive advantage in themselves.

The following section is devoted to examining how the personnel function carries its role in leading Japanese organizations. The various activities which will be examined include training, career development, empowerment, the decision-making process and remuneration.

Training

On-the-job training (OJT) is the formal process of developing employee skills – it is often attributed as a task to managers and supervisors, who are expected to spend as much as 20–30% of their time training their staff and subordinates. In fact in Japan, management effectiveness is reflected by the effectiveness of employees.

On-the-job training is a low cost approach and tends to be very company specific. The reason it works successfully in Japan is due to the wide practice of industrial paternalism (Ozawa, 1982). According to Ozawa (1982) the **industrial paternalistic approach** leads to a wide variety of benefits which include:

- employee loyalty, therefore encouraging employers to invest in people;
- market regulation in terms of skills provision and availability, therefore reducing competition for skills and the poaching of experienced labour;
- benefits of mutual interest to the individual in terms of pay and rewards and to the company in terms of high productivity levels;
- growth for both the individual and the company go hand in hand;

- a clear career path reducing suspicions about vacancies and how they are filled; and
- generous sharing of skills and knowledge.

In relation to this concept of sharing skills, Ozawa (1982) writes:

> Not only life employment but the seniority system as well enhances the effectiveness of on-the-job training. Senior workers are willing to impart their accumulated knowledge and experience to juniors, particularly to new recruits, without fear of jeopardizing their own positions through possible future competition. The system leads to an effective interpersonal way of imparting knowledge down through the hierarchy, and training thus becomes built in.

Job rotation is an informal style of training employees. It is used for a variety of reasons, including the following:

- to enable employees to develop all round, be familiar with all aspects and to have knowledge of all activities in their organization;
- to enable managers to interact and learn how to communicate with various people within their organization; and
- to enable managers/workers to be 'ambassadors' for their organization externally.

Career progression in management

All new recruits are expected to start at the bottom of the organization before they can assume managerial responsibilities. They have to develop a capacity to understand wider business issues and as such, experience supercedes knowledge. In Japan, management is not considered as a profession but is relative to status and seniority levels.

OJT and job rotation are the essential means for developing managers and for employee promotions and movements. Unlike in the West, the Japanese rely on a **push leadership style** rather than a pull style (Fig. 12.11). Leaders are people who have thorough understanding of all business activities and have undertaken assignments in nearly all the areas of business operations. It is therefore from a good knowledge base and thorough understanding that decisions are made.

In Japan, the criteria for promoting people to high positions are based on a series of observed practices and guidelines.

- Employees are not promoted because of their high level of education. In many cases most employees tend to have an adequate degree of knowledge and education. Experience will however spread the aspirations of individuals and the speed of progression will vary according to ambition, achievement and assimilation of knowledge.
- Attitude towards promotion is that it is part of a long-term career progression plan. Individuals do not consider promotion in absolute

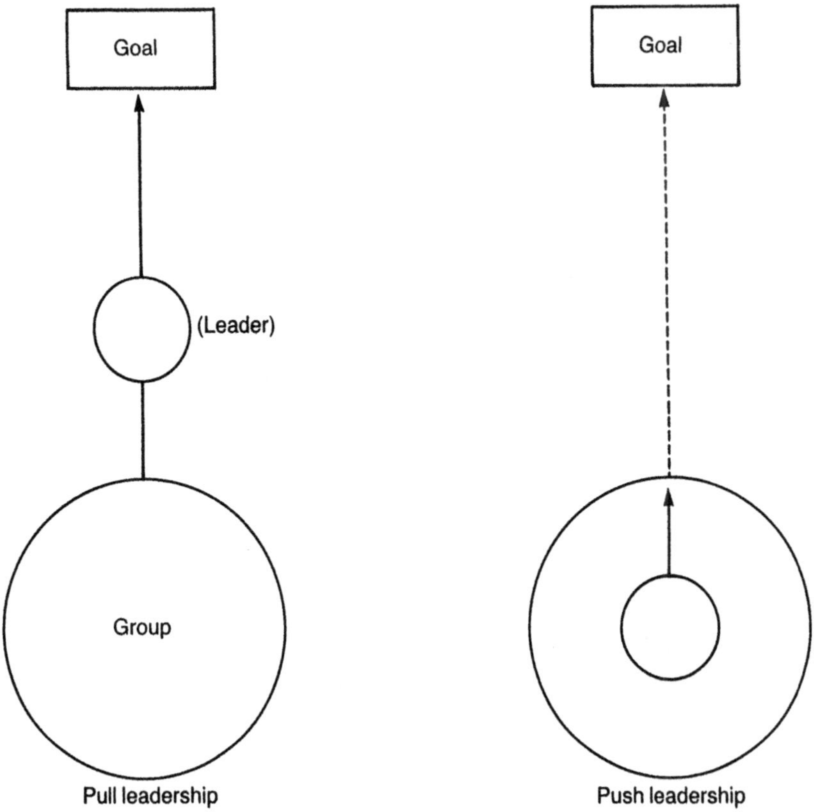

Figure 12.11 Pull and push leadership styles. *Source*: Ballon (1992).

terms as a position achieved; rather they view it as another step upwards towards enhancing their career profile.

- Knowledge is not the deciding factor in choosing people to manage resources but competence tends to be based on experience and wisdom is developed.
- In order to enhance corporate standards of knowledge and competence, the Japanese, in promoting an individual, will not look at the exceptional cases but rather the organizational norm and average.

Empowerment in Japan

Unlike Western practices, in Japan, the decision-making process is not dependent on corporate policy and objectives. It is very much a case of a problem/situation which needs to be acted upon, particularly if it impinges on organizational performance. Then it is left to the appropriate levels to act.

In Japan, the decision-making process is not just a case of observing rules and procedures and their implementation but rather the deployment of knowledge and skill levels for enhancing organizational performance.

People are appreciated for their knowledge and experience and therefore there is total delegation of operational problem solving at the local level. The reason managers can entrust employees to make decisions at their level is because there is a safeguard which is the team, and all individuals are part of the team. So if an individual makes a decision, it will be because his/her team has empowered him/her to act on behalf of the team.

Empowerment means that the concept of volunteered accountability can be made to work, since the communication of information is not in a hierarchical way, the sharing of information is part of the culture of full trust, common organizational goal, problem-solving, etc.

Employment relations

The concept of life-time employment in Japan has been misunderstood. It does not necessarily mean employment for life, as explained by Inohara (1990):

> This may be easily misunderstood as if it were institutionalized as such . . . however, long-term employment is not always possible . . . therefore, long-term employment is, in fact, a social norm to observe and an ideal to pursue.

Japanese prefer to speak about regular employment, which is not permanent. They strive to retain people, however, through various means, such as:

- continuous training;
- long-term career path;
- respect for people;
- interdependence of organizational growth and employee growth;
- performance improvement linked to peoples' skills and knowledge availability; and
- programmes of OJT, job rotation and external secondments.

Employee remuneration

The Japanese use a principle called *Nenko* which basically means seniority – and recognizes the fact that people grow and become mature, wiser and technically better. This means that the quality of contribution becomes better. Salaries are given accordingly, rewarding number of years' service rather than employee competence and ability.

12.5 Benchmarking customer–consumer satisfaction

It has been suggested that there are three different types of customers (Lawton, 1991):

1. **end users** use the product or service provided;
2. **brokers** transfer the product/service to the end user; and

3. **fixers** transfer, repair or adjust the product/service for the benefit of the end user.

There is a need to develop strategies for building long-term relationships with each and ensuring their total satisfaction all the time. The identification of key *drivers* for customer satisfaction becomes therefore a task of paramount importance, as the following quotation suggests (Miller, 1991):

> Measuring customer satisfaction is an important issue: if you don't measure customers' satisfaction, you can't identify customer trends in the long term and won't be able to react in a timely fashion.
>
> (R. Kreiner, Marketing Director, McDonald's, Germany)

Customer satisfaction can be measured in a wide variety of ways (Miller 1991):

- **Holiday Inn** and **IKEA** provide questionnaires to be filled in on the spot by customers and dropped into special boxes;
- **Dell Computers** includes a questionnaire diskette with delivered computers, and thanks its customers/respondents with ten free diskettes;
- **VW** conduct a survey of approximately 660 000 customers in Germany every two years;
- **Compaq** use a combination of telephone and written surveys;
- **Panasonic** include reply cards in each product sold;
- **Domino's Pizza** use 10 000 unidentified customers who receive $60 every year to buy 12 pizzas and are then asked to evaluate the quality and service experienced;
- **Bang & Olufsen** extends its guarantee for another year to customers who respond to their questionnaires.

The following benchmarking frameworks can be used to measure customer/consumer satisfaction.

Customer service audit framework

This model, proposed by Professor M. Christopher of Cranfield School of Management and Mr Richard Yallop, managing director of Customer Service International Ltd, has four steps:

1. identifying the aspects of the service most highly rated by the customer, e.g.
   - frequency of delivery
   - time from order to delivery
   - reliability of delivery
   - orders filled completely
   - accuracy of invoices
   - credit terms offered;
2. prioritizing using a simple ranking system or the 'trade off' technique;

3. benchmarking against key competitors, and then, using the outcomes of steps 2. and 3., drawing up a complete service profile for individual suppliers against key competitors (Fig. 12.12); and
4. measuring service performance against customer priorities.

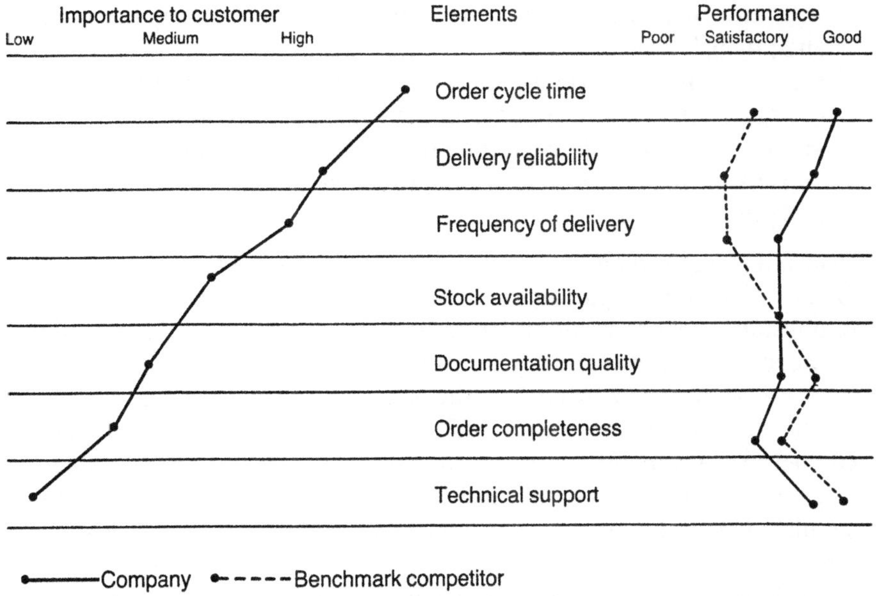

Figure 12.12 **Customer service complete profile.**

The customer–consumer model at P&G

Procter and Gamble (Soap division) have aligned their process to meet the needs of their customers (trade sector) and the ultimate consumers (households) (Saxton and Locander, 1991). Four key processes were identified including:

- Sales
- Advertising
- Product supply
- Product development

The key elements of service identified by customers and consumers of P&G products include the following:

1. **performance** benefits provided by the product;
2. **dependability** i.e. consistency in performance;
3. **price** i.e. what consumer likes to pay;
4. **availability** i.e. ease of obtaining product;
5. **awareness** of the product and its benefits;
6. **image** benefits concerned with emotional needs; and
7. **service** i.e. how the product is delivered and followed up.

Figure 12.13 illustrates the model that P&G (Soap division) uses to ensure total customer/consumer satisfaction.

- **Measurements of enablers** The outside dotted lines are those measures of key process performance essential for satisfying customer/consumer needs. For example, for **consumers** the advertising process is concerned with advertising, promotion, development of packaging, the concept and pricing.
- **Measurements of outcomes** Each critical process needs to impact on the customer/consumer. Specific measures are used for each. For example, in the context of **customers**, the process concerned with product supply impacts on the trade sector through dependability, availability, image and price.

Benchmarking customer–consumer satisfaction with QFD
Quality function deployment (QFD) represents a powerful tool for establishing customer–consumer needs and the degree of compatibility between what suppliers do and how satisfied customers/consumers are.

Customer wants or 'whats' can be established using the core aspects of the product or service and the additional features that give suppliers a competitive advantage, and tend to establish differentiation. Using the Kano model (Fig. 12.14) all the wants or needs can be established and circulated to the 'hows' (i.e. supplier activity that translate them into tangible outcomes).

QFD allows competitive assessment. Suppliers can benchmark their ability to deliver each customer 'what' against competitors. It also allows them to establish how well the customers rate supplier performance.

In addition, suppliers have the opportunity to benchmark technically 'how' they translate customer wants against their key competitors.

12.6 Benchmarking in marketing

Marketing is the key link between what an organization does/should do and the outside world, including customers and consumers, existing and prospective ones. Marketing has evolved over the years, to reflect changes in the marketplace but also to realign with internal organizational initiatives. Whilst it is a key driver to business performance, it is however recognized that today, for marketing to be effective, it has to be integrated with other functions.

Marketing, as a concept, has always centred on 4Ps (products, places, prices and promotions). However, for marketing to be effective in the 1990s, there has to be close attention given to a fifth P, the people process.

- People are key to service delivery, quality and relationships with customers.

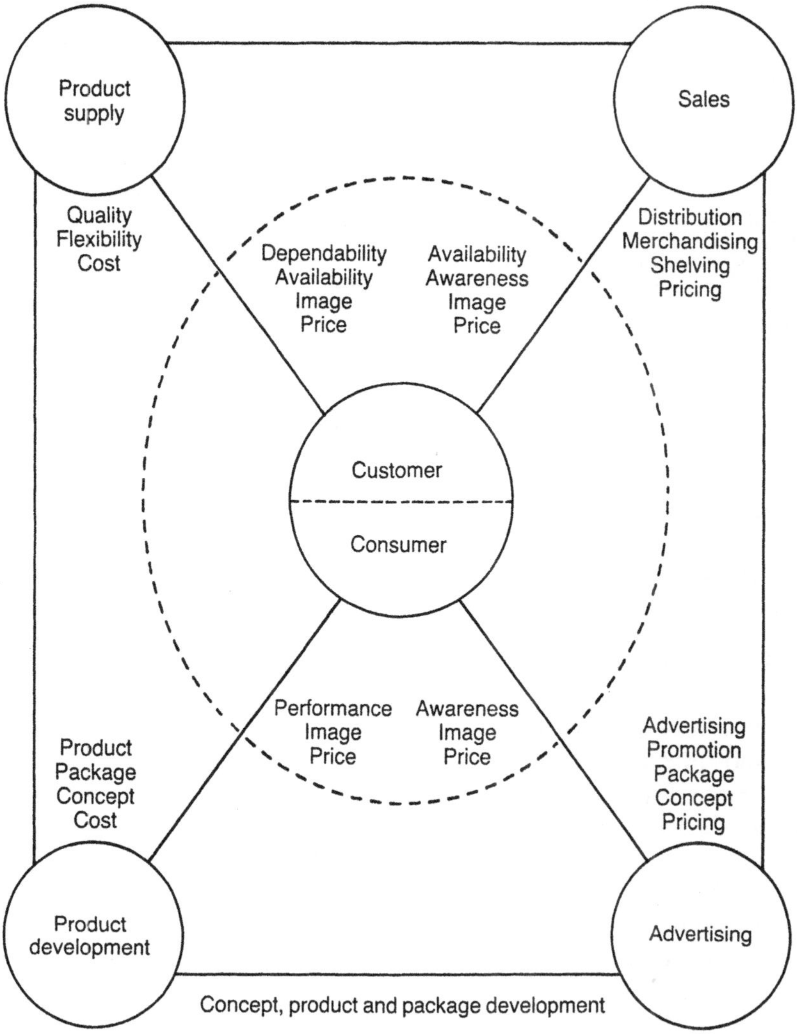

Figure 12.13 Customer–consumer model at P&G.

- If people do not believe in the product/service, nor will the customer – people have to be managed properly.
- Managing products and services and managing people go hand in hand.
- People impact on:
  - positioning,
  - pricing,
  - distribution,
  - reputation, and
  - sales.
- People in marketing have to adapt constantly to change.

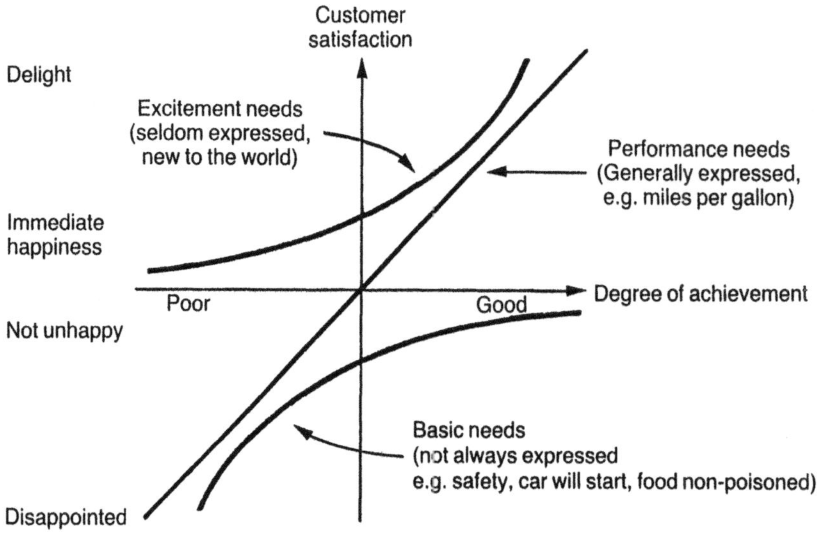

Figure 12.14 Determining customer/consumer wants with Kano model.

- Critical elements of modern marketing are people related:
  - flair,
  - interest/understanding,
  - innovativeness, and
  - communication.
- People need to be nurtured and trained in:
  - research,
  - creation,
  - distribution, and
  - serving the customer.

Marketing in the 1990s relies more and more on responsiveness, alertness, closeness to the customer, creativity, good communications, risk taking, and the exploitation of levels of synergy from within the organization.

Marketing concerns itself with the customer/consumer. It has the primary role of translating needs and wants into innovative activity that generates profits. Marketing must 'point the torch' at the customer and help point the way to go.

Marketing works in two ways:

- developing new products and services capable of fulfilling unmet customer needs; and
- building customer loyalty for repeat purchases through the establishment of brand names.

Development of new products/services
Most products and services in the 1990s have shorter life cycles and as a result of customer/consumer sophistication and changing perceptions, there is a need to continuously innovate. The successful formula is to:

- invest,
- innovate, and
- improve.

True differentiation can only come from radical innovation.

The management of brands
Brands and products are not necessarily the same thing. However, they are both interdependent and most organizations try to sustain the performance of one with the other. As Stephen King (WPP Group, London) describes:

> A product is something that is made in a factory; a brand is something bought by a customer. A product can be copied by a competitor; a brand is unique. A product can be quickly outdated; a successful brand is timeless.

The role of marketing, therefore, is to sustain customer loyalty through strengthening brand names, not necessarily through incremental innovation on existing lines but radical innovation that leads to differentiation and a commanding place in the market place.

Effective marketing means creating a strong customer base through the establishment of powerful brand names, and protecting that customer base through continuous innovation.

12.6.1 Marketing in Japan

The Japanese have always been spoilt for choice. The situation is, however, changing. The Americans are, for example, closing the gap. Ford has replaced Honda as best selling car maker in the US. Japanese PC makers are being undercut by American low cost producers.

In Japan, there are approximately 13 retailers for every 1000 inhabitants compared with six per 1000 inhabitants in the US. To maintain the effectiveness of distribution networks, the Japanese use regular visits rather than brand building or studies in changing consumer behaviour. Product proliferation in Japan is, however, coming down gradually because of fast erosions in market share and poor performance of many product lines. Techniques such as market research, brand building and test marketing, are increasingly being relied upon.

The Japanese will, however, maintain a keen interest in improving quality of existing products and services, and quality rather than image will remain the touchstone of Japanese companies.

12.6.2 Benchmarking and marketing – what is the link?

Modern competitive strategies are designed to focus on the end customer and their effectiveness can be measured in customer satisfaction, rather than in profitability increases – as was traditionally the case. It is only by integrating the voice of the customer through a clear understanding of their needs, with the voice of the process (the organization's capability to deliver) that success in the marketplace can be achieved.

Modern marketing has to be an integral part of competitive strategies (Fig 12.15). As an integral component of competitive strategy, modern marketing can be described as a process that focuses on 3Cs (Fig 12.16).

1. **Customer** i.e. the establishment of customer satisfaction through delivery of innovation capable of fulfilling needs; the assessment of future needs and unfulfilled needs.
2. **Company** i.e. helping to develop customer care processes and activity which focus entirely on the end customer.

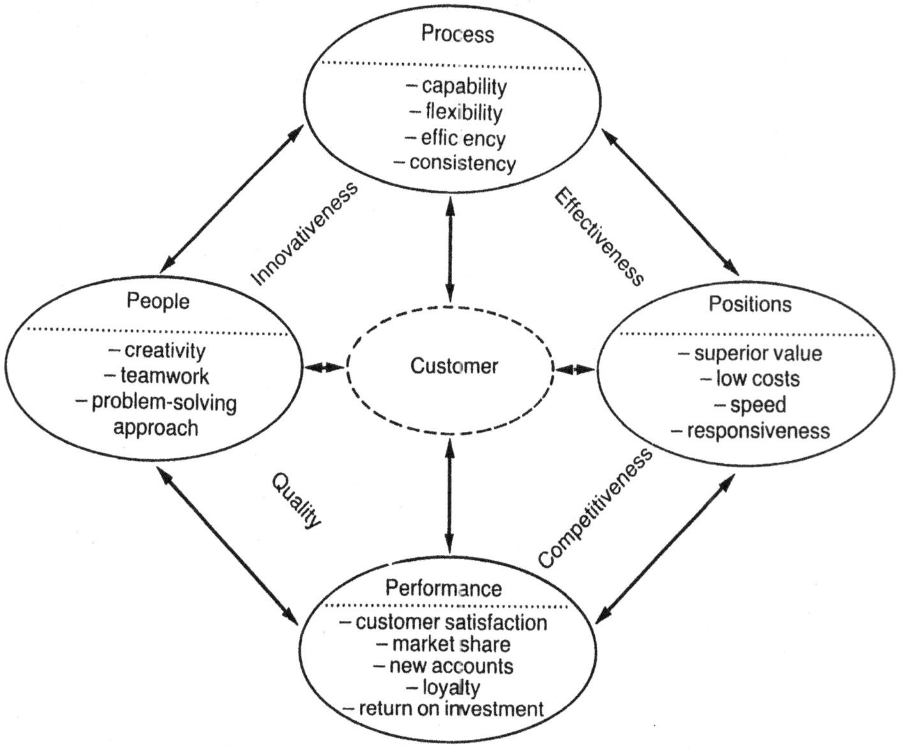

Figure 12.15 Modern marketing – an integral part of competitive strategy.

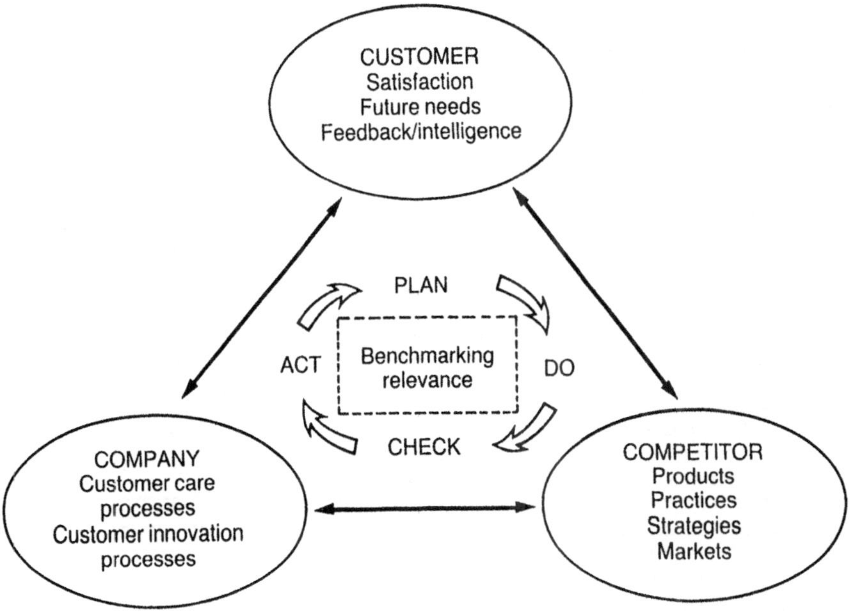

Figure 12.16 The proactive role of marketing.

3. **Competitor** i.e. gathering intelligence on competitor moves, performance of their products and services and their impact on the marketplace.

Benchmarking is a powerful tool that optimizes marketing effectiveness under the above three headings. It helps organizations focus on processes and establish high standards of effectiveness. It drives marketing to keep focusing on the 3Cs and using PDCA eliminates any risk of complacency (Fig. 12.17)

The following examples illustrate various ways in which benchmarking can be applied in marketing.

Benchmarking the power of brands

A benchmarking study was conducted by Landor Associates who developed a tool to measure the power of brand names (Aaker, 1991; Ryan, 1988). The instrument that was used on 1000 American consumers was to measure the following:

- share of mind score – to measure brand recognition;
- esteem – a measure of favourable opinion that people have for companies and brands they know.

Ratings obtained from the survey were then averaged to form an overall 'image power' score. Table 12.5 illustrates the most powerful brands in the US. The Landor study demonstrated that high recognition scores in

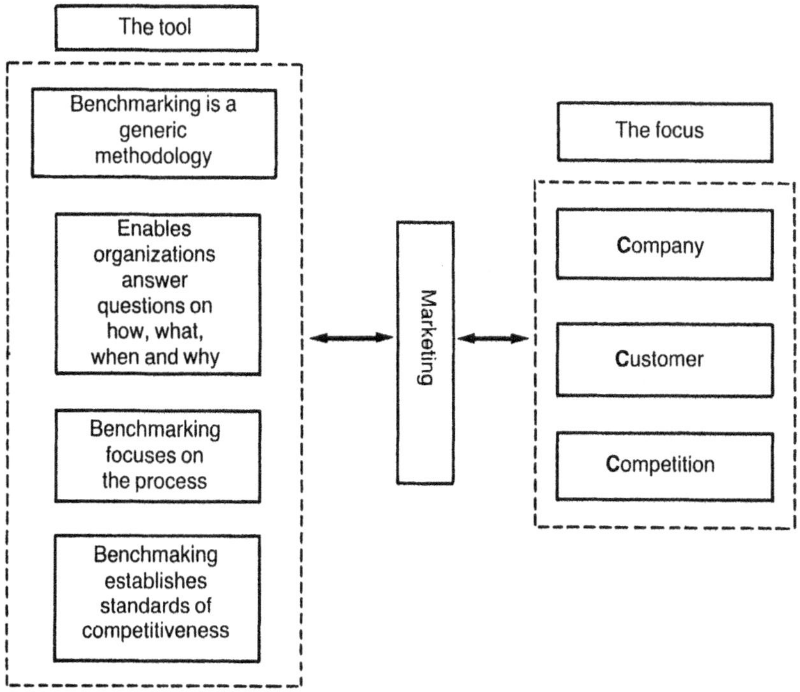

Figure 12.17 **Benchmarking – a key driver for marketing.**

brands lead to successful launches of brand extensions and, in some instances, with little or no marketing support. For example, in the mid 1980s, the Coke name was attached to nine different soft drinks, including Diet Cherry Coke, a brand that received no advertising support at all (Aaker, 1991).

Benchmarking the leading brands, 1925 and 1985

This study was conducted by the Boston Consulting Group looking at leading brands in 1925 with those of 1985, by focusing on 22 product categories. As Table 12.6 illustrates, in 19 categories the leaders remained the same (Aaker, 1991; Wurster, 1987).

The Boston study demonstrates that high recognition in brands can be considered as a powerful asset. Sustaining the power of brands through repeated exposure means that entry becomes almost impossible, since even high quality products/services backed by huge advertising budgets will find it difficult to challenge opinions of customers.

Benchmarking market share

Market share is the ultimate way to measure the performance of brands. Table 12.7 shows an example of brand performance over a four-year period in the haircare sector.

Table 12.5 Most powerful brands in the US

Image Power Rank Order	Company/Brand	Esteem Index	Share of Mind Index
1	Coca-Cola	68	78
2	Campbell's	67	60
3	Pepsi-Cola	61	67
4	AT&T	64	63
5	McDonald's	50	77
6	American Express	60	65
7	Kellogg's	58	64
8	IBM	65	58
9	Levi's	63	58
10	Sears	59	62
30	Rolls-Royce	63	46
169	Nissan	43	66
177	Datsun	41	67
667	Asahi	28	27

Benchmarking advertising spend

Reaching customers and consumers is measured through advertising spend and sales performance. In the advertising spend by the top 100 US companies in 1992, information technology companies are found to lead the pack. The marketing tools most frequently used by info-tech companies include:

- advertisement in trade publications (39%);
- trade shows (19%);
- direct mail marketing (19%);
- person to person sales (6%);
- non trade business publications (6%);
- television (6%);
- telemarketing (6%);
- public relations (5%);
- advertisement in consumer publications (2%);
- consumer targeted publications (2%);
- radio (2%);
- fax marketing (2%);
- cooperative ads with dealers (2%).

Examples of marketing communication strategies in leading companies

Hewlett Packard Highly committed to direct marketing, market research and broadcast advertising. Established a direct marketing council and a task force to link various customer bases.

Table 12.6 Leading brands period: 1925–85

Product	Leading brand 1925	Current position 1985
Bacon	Swift	Leader
Batteries	Ever Ready	Leader
Biscuits	Nabisco	Leader
Breakfast cereal	Kellogg	Leader
Cameras	Kodak	Leader
Canned fruit	Del Monte	Leader
Chewing gum	Wrigley	Leader
Chocolates	Hershey	No. 2
Flour	Gold Medal	Leader
Mint candies	Life Savers	Leader
Paint	Sherwin–Williams	Leader
Pipe tobacco	Prince Albert	Leader
Razors	Gillette	Leader
Sewing machines	Singer	Leader
Shirts	Manhattan	No. 5
Shortening	Crisco	Leader
Soap	Ivory	Leader
Soft drinks	Coca-Cola	Leader
Soup	Campbell	Leader
Tea	Lipton	Leader
Tyres	Goodyear	Leader
Toothpaste	Colgate	No. 2

Table 12.7 Benchmarking in haircare sector (1988–91) – brand performance

Company	Brand	1988	1989	1990	1991
Procter & Gamble	Head & Shoulders	12	10	11	13
	Wash'n go	–	–	21	23
Elida Gibbs	Timotei	10	9	8	7
	Dimension	5	4	2	4
	All Clear	4	3	–	–
Beecham	Vosene	6	5	4	4
	Silvikrin	3	3	2	2
Alberto C	VO5	3	4	–	–
Gillette	Silkience	3	3	–	–
Other Brands		40	40	36	30
Boots' own label		–	–	4	5
Own labels		12	14	12	12

Source: Trade Estimates and Market Assessment

IBM Staged a major reorganization to try to regain its competitive advantage. Uses aggressive trade show programme that includes more than 200 expositions per year worldwide.

Microsoft Direct marketing and launch of newsletters for customers and resellers.

Digital Sponsorships for sport on TV. Desktop direct – customers can order PCs directly over the telephone.

Toshiba Splits its budget evenly between trade and general business publications, using the latter for building Toshiba's brand name.

Kodak Focuses its communications on the influencers for its business-to-business products.

Fujitsu Revised its marketing strategy to accompany its new emphasis on value added resellers.

Texas Instruments Is placing more emphasis on its corporate identity through new packaging for products sold in office product superstores. The company has adopted the criteria of the Malcolm Baldrige National Quality Award, and is reengineering all its key processes including marketing communications, with the view of reducing cycle time.

3M Has developed unified messages in common markets to increase the power of its marketing messages and save money at the same time.

Dupont Relies heavily on direct marketing and telemarketing. This means that field salespeople are not required in great number allowing Dupont to have total closed loop sales.

Unysis Is using a campaign launched in mid 1991 with the slogan 'we make it happen' using case histories highlighting customer solutions, pointing to its high profile customers.

Walt Disney Euro Disney uses direct mail to invite companies with 50 employees or more to have free membership, allowing employees of member companies to have access to many benefits, including reduced fees at the theme park. Euro Disney aims to establish long-term strategic marketing partnerships with companies.

12.6.3 Best practice marketing

This section includes examples of best practice marketing in a number of leading companies.

- **Digital**
  - Of Digital's revenues 41% come from servicing customers.
  - Digital addresses customer needs through account groups.
  - Every customer has a Digital team devoted to his service.
  - Digital are experimenting with a concept making each customer account a business unit with an appointed manager responsible for customer satisfaction and profitability.

- Digital relies on IT networks for providing information to its various employees who service customer accounts.

- **Nabisco Foods**
  - Nabisco relies 100% on IT to process customer orders on time intact, accurately priced, proper billing and promotional discounts on products.
  - Nabsico meets promised delivery time at 99% in comparison with 94.1% industry average.
  - Nabisco receives 88% of customer orders electronically.
  - Nabisco uses electronic data interchange (EDI) and is pioneering electronic order submission.
- **L L Bean**
  - The golden rule at L L Bean is to 'sell good merchandise at reasonable profit, treat your customers like human beings, and they will always come back for more'.
  - In the early 1980s on-line customer service and toll-free phone numbers were introduced.
  - Now telephone representatives handle over 10 000 customer enquiries a day.
  - Employees are taught to follow the golden rule and are empowered to act to ensure customer satisfaction.
- **Elanco** (Animal Health Division of Eli Lilly & Co)
  - Elanco learnt how to stop relying on pushing products out and bringing market in.
  - Through the use of market research, Elanco learnt that it is better to ask than assume and it is more important to listen than act.
  - As explained by Elanco's European Sales and Marketing Director, 'We have had to stop talking purely about our products, not easy for a research driven, high tech business like pharmaceuticals – and get customer focused'.
- **ICI**
  - ICI created a market focus bureau to create a better market-focused company.
  - The bureau runs the experience databank, which collects successful practices and strategies from around the world.
  - When units need information the databank provides reports, presentations and articles with contact reference for follow up.

12.7 Benchmarking in education

Changes in the business community are having significant impact on employee development and training and the recruitment of graduates with knowledge and skills that deal with current business issues. It is almost impossible to ignore the scale of changes taking place in the business world and academic institutions have to rise to the challenge. As explained by Sink (1993):

> Clearly higher education is on the precipice of change. There are those who cling to the 'good ole days', those who are early adopters but are not quite sure what they are adopting, and those who are sceptics or at least cautious observers of the situation. . . We, in higher education, are uncomfortable using terms like 'customers', 'value added', 'processes', and 'systems' because we feel they are applicable only for business and industry. Or, perhaps, is it really because we resist the very concepts themselves and hide behind the excuse of language?

12.7.1 TQM in education – is there a need?

To answer this question perhaps let us refer to an academic institution that has used TQM and benefited from it. Oregon State University (OSU) in the US was facing the prospect of having nearly $40 million in state funding cut from its budget. TQM was therefore adopted through desperation and an urgent need to survive. After a short period of piloting, Oregon State University started the implementation process through top management involvement, strategic planning, defining customer needs and key processes. TQM has succeeded at OSU and the following factors were found to be critical for its success in an academic setting (Coate and Maser, 1993):

1. support from the top;
2. just do it;
3. teams are everything;
4. you need a champion;
5. breakthrough planning (focus on key processes);
6. try the service side first (before academic delivery).

12.7.2 Benchmarking in academia

Like TQM, benchmarking is a relatively new concept to academic institutions. However, students and employers are more and more looking for value for money and quality delivery in terms of skills and knowledge.

The following are three benchmarking examples representing academic institutions:

Example A

To represent performance of students during and after university life in the UK, the following measures were used in this benchmarking exercise:

1. **permanent employment** Percentage of graduates in permanent employment after six months of graduation (1990/91);

2. **unemployed graduates** Percentage after 6 months of graduation (1990/91);
3. **research or further study** Percentage doing research or further study (1990/91);
4. **first Class Honours** Percentage winning a first class degree (1990/91).
5. **completion rates** Percentage successfully completed first degrees (1990/91);
6. **graduate students** Number of students on course which normally requires a first degree for entry in 1991/92;
7. **international students** Percentage of all 1991/92 students including exchange students who will graduate from a non UK university.

Example B

This benchmarking exercise looked at various critical processes within academic institutions in the UK. These core processes represent the overall strength of each academic institution, its credibility and ability to compete in attracting students on courses and successfully bidding for research money.

1. **Entry requirements** i.e. entry qualifications of students beginning degree courses in 1991, weighted to reflect difference between offer and achieved entry levels.
2. **Student/staff ratios** The average 1991/92 student/staff ratios across the institution, based on full-time equivalents.
3. **Staff with PhDs** Percentage of academics with doctorates in 1991/92 (excludes academic-related staff with no teaching load).
4. **Professionally qualified staff** in 1991/92.
5. **Library spending**/expenditure on books, periodicals and staff.
6. **Dual funded research income** i.e. total 1991/92 research income from research councils and medical charities per full time academic staff.
7. **Contract research** income during 1991/92 from all other sources per full time academic staff.
8. **Student accommodation** of all students where accommodation is provided by institution.

Example C

This ranks universities in the UK according to subject strength. It is argued that most universities will have pockets of excellence in some subjects but may not necessarily be good at everything. Aggregate scores may not be useful because they will determine pecking order and not necessarily reputations in relation to specific topics.

Unlike published statistics, this benchmarking example is thought to be new.

- It ranks universities according to seven main areas focusing on teaching capability rather than research strength.

- It has to be updated once per academic year to reflect consistency and show trends of improvements.
- Rankings are based on entry grades, staffing levels and departmental budgets. In this way, the rankings will reflect reputation and demand for places, as well as the resources available in the main subject areas.

12.7.3 Best practice in education

Oregon State University (OSU)
Using cross-functional teams for problem solving has led to some of the following benefits (Coate and Maser, 1993):

- **physical plant team remodelling** reduced average duration of remodelling jobs by 23% with estimated cost savings of $219 539;
- **budgets and planning team** improved budget-status-at-a-glance report to meet customer needs and cut preparation time by 50% with estimated cost savings of $3600; and
- **human resources staff benefits team** increased number of phone calls getting an initial human response by 83% with estimated cost savings of $6667.

Georgia Institute of Technology (GIT)
GIT implemented a continuous quality improvement (CQI) programme in a very systematic way.

1. They had the **vision** to become the premier technological university in the nation.
2. Their **goal** was to target not only **what** was taught but **how** it was taught from the perspective of their customers' needs.
3. **Processes** were established with the view of:
   - examining the impact of alternative evaluation and grading procedures;
   - timing of material introduction; and
   - alternative delivery methods.
4. **Measures** were created to monitor their effectiveness in meeting customer expectations.
5. They **developed** data collection strategies (Ammons and Gilmour, 1993).

Paul D Camp Community College (PDCCC)
In the mid-1980s PDCCC suffered big drops in student enrolments (between 1984–85 to 1988–89 headcount dropped from −8.2% to −20.3%) from previous years and full-time equivalent students (FTES) went from −13.8% to −8.4%. The college also received one of four unsatisfactory ratings among public colleges and universities in Virginia for its state mandated programme of student outcomes assessment (Joyer, 1991).

This therefore called for big changes, and coincided with the arrival of a new college principal. A new philosophy of management was introduced, based on the book *In Search of Excellence* by Peters and Waterman. Some of these principles included:

- assessing strengths and weaknesses, and developing a mission to meet customer needs;
- a new managerial approach, with the development of goals, objectives and strategies to reach them;
- emphasis on teamwork, problem solving and a positive attitude where risk taking is widely encouraged through active participation and staff empowerment;
- better communication through the establishment of a daily bulletin and the active promotion of the college within the wider community;
- a concrete plan for conducting continuous improvement with specific targets and milestones;
- the use of measures such as:
  - surveys to measure student satisfaction;
  - surveys to measure employer satisfaction with college graduates;
  - student enrolment as compared to institutional enrolment projections; and
  - data collection to assess student retention rates.

Amongst the various benefits achieved since the introduction of TQM at PDCCC are:

- PDCCC moved from being bottom to top out of 213 community colleges in Virginia;
- full-time equivalency enrolment increased 30.8%;
- headcount enrolment increased by 40.4%;
- student retention rate increased by 25%;
- faculty class sizes increased, reflecting increased institutional productivity from 0.88% to 0.92%;
- listening and responding to needs of students resulted in the development of **self directed access to new delivery systems** (SANDS) (Joyer, 1991).

12.8 Benchmarking in healthcare

There has been significant growth in the utilization of TQM principles in healthcare, including the use of benchmarking in recent years. In the US, for instance, changes in healthcare called for modern management principles and a need to increase quality and productivity levels. The changes which took place in the US included a rapid growth in healthcare expenditure from around $42 billion (5.9% of GNP) in 1965 to very nearly $700 billion (12% of GNP) in 1990. The projection is that healthcare would account for around 20% of GNP by the year 2000 (Umachonu, 1991).

In the US, patients tend to be much more involved in the healthcare purchase than their counterparts in Europe. So, hospitals compete with each other on outcomes, standards, prices and quality which is considered as a marketable commodity in the US. In addition, the cost of malpractice litigation is constantly rising and there is pressure from government and healthcare bodies, in particular from the Joint Committee on Accreditation of Healthcare Organization (JCAHO) and the American Group Practice Association, to deliver a high quality low cost service (Fortune, 1989; Feeney, 1991; Koska, 1991).

Quality management and the use of benchmarking are growing practices within the National Health Service (NHS) in the UK. Many reforms have taken place in recent years. In particular the publication of the Griffiths Report (Griffiths, 1983) highlighted the lack of attention paid to quality assurance and consumer satisfaction:

> . . . there is little measurement of health output; clinical evaluation of particular practices is by no means common, and economic evaluation of those practices extremely rare . . . whether the NHS is meeting the needs of the patient and the community, and can prove it is doing so, is open to question.

In the past, quality control was established nationally in scientific departments such as pathology and radiology. Quality of clinical care, however, can only become a management issue as a result of external pressures and demands for better quality assurance, from increased patient expectations, consumer groups and rising levels of medical negligence litigation (Feeney, 1991).

In 1984, Britain, in common with other European Countries, committed itself to a World Health Organization declaration to develop 'effective mechanisms for ensuring quality of patient care within the health system' by 1990. In 1989, The Department of Health began its present reform of the structure of the NHS and published its white paper *Working for Patients* (Gray, 1989). The reforms stressed the importance of quality assurance by creating a purchase–provider split and an informal market which competes on price and quality. In addition the reorganization of the NHS let to the creation of Trusts and contracts for health services based on specifications of quality standards and regular measurements of performance and delivery in all aspects.

The compulsory medical audit was introduced as 'the systematic critical analysis of the quality of medical care'. It examines:

- procedures used for diagnosis and treatment;
- use of resources;
- resulting outcome and quality of life for the patient.

Publication of *The Health of the Nation* document set performance standards and targets for the future. In it the benchmarking framework is set out. It is to be implemented nationally as follows:

- select five key areas for action;
- set rational objectives and targets in the key areas;
- indicate the action needed to achieve the targets;
- outline initiatives to help implement the strategy; and
- set the framework for monitoring, development and review.

The key areas chosen include:

1. coronary heart disease and strokes;
2. cancers;
3. mental illness;
4. HIV/AIDS and sexual health; and
5. accidents.

12.8.1 Benchmarking the benchmark in healthcare

The Health of the Nation sets clear targets with clear dates:

- reducing death rates for both coronary heart disease (CHD) and strokes in people under 65 by at least 40% by the year 2000;
- reducing death rate for breast cancer in the population by at least 25% by the year 2000;
- reducing the death rate for lung cancer by at least 30% in men under 75 and 15% in women under 75 by 2010;
- to reduce the overall suicide rate by at least 15% by the year 2000;
- to reduce the death rate for accidents among children aged 15 by at least 33% by 2005, young people aged 15–24 by at least 25% by 2005.

This benchmarking initiative is perhaps one of the most radical and daring plans yet to be introduced in the NHS. It is intended to change the culture of healthcare provision by having a clear vision, key priorities, confidence and know-how, better focus and by setting priorities from a patient perspective. This new culture will also be one where targets become only a relative milestone and where the key drive is never-ending improvement. Benchmarking in the future will become an integral element of health care provision at all levels. As stated in *The Health of the Nation*.

> Improving overall health is not just a new management priority. It must become the central concern of all sectors of the NHS and should be integral to the work of every health professional in the NHS.

12.8.2 The health service indicator set – a benchmarking methodology

This basically represents comparative information in various areas, using performance indicators. Information has been published since 1983. Since all the processes in the NHS are very similar, publication of sets of performance indicators enables local health care managers conduct comparisons, prepare action plans and identify who the best performers are and where the best practice is.

Although still at development stage the Health Service Indicator set (HSI) is extremely useful. It includes key indicators mentioned in *The Health of the Nation*. It focuses purely and simply on quantitative performance and assessment of services.

The HSI is not a final tool for making major decisions. Although it presents useful benchmarking information, its application at local level requires some work.

- In areas where the quality of service indicators are still lacking, users are encouraged to qualify analyses of performance using alternative information.
- Indicators are unable to demonstrate what has actually happened, why the service is delivered in a particular way and how it is delivered. Users have to translate the significance of each indicator by mapping the process concerned, understanding behaviour, analysing the effectiveness of the enabling practices and then put outcomes/measures in the right context.
- The HSI is a retrospective compilation of performance standards. It does not reflect day-to-day performance management and only points at trends and changes over time.

Other benchmarking frameworks applied in the NHS include:

- CHKS Database
  As part of the CASPE projects, a database has been set up for Trusts, providing national benchmarks market intelligence with providers of information remaining anonymous;
- NHS management executives are considering a **tool box** that can look at a whole range of indicators covering various processes:
  - health policy
  - resources
  - activity
  - quality
  - efficiency and effectiveness.
- Patient charter standards
  Rights and standards are published in the Patient Charter document. Monitoring of performance of all providers will be done centrally. A

report based on a set of six key indicators will be published in July 1994, representing a league table of performance for the period 1993–94. The league tables to be published in 1994 will cover the following areas:
  - immediate assessment in accident and emergency;
  - thirty-minute wait in outpatient clinics;
  - cancelled operation standard;
  - waiting time for treatment in a number of key specialities;
  - information on day surgery procedures; and
  - ambulance emergency response times.
- Nursing and Practice Development Units
  An accreditation system is used as a standard for these units, which is based on quality, innovative practice, etc. Those who apply for accreditation at Stage 2 have to fulfil standards under 18 different criteria.

12.8.3 Examples of best practice in healthcare

Good practice in Healthcare Trusts in the UK started to be publicized in a newsletter called *Regional Charter Round Up*. The following are examples of best practices extracted from the newsletter.

- Full district nursing records are kept at the patients home (South Cumbria Community and Mental Health Unit).
- Performance improvement in communication with patients joining waiting lists ensures that letters advising of the wait are sent immediately (Newcastle General Hospital).
- Trust board has established a patients forum to provide objective assessment and comment on patient services (Newcastle RUI).
- Improvements made to facilities for the disabled – wheelchair clinic includes sign posting, beverages, access parking and toilet facilities (Durham Community Health Services Unit).
- An annual Quality Award for initiative judged as best quality project for the year (Southport and Formby NHS Trust).
- A multidisciplinary group set up to examine monitoring methods for quality standards including the Patient's Charter to ensure valid and consistent methodology (Queen Elizabeth Hospital).
- New complaints procedure: the complainant receives a response within 24 hours of receipt, is informed of progress after seven days and the investigation is completed within 28 days (Sandwell Hospital).
- A newsletter for patients on waiting lists to provide information about the hospital and appropriate support groups (Papworth Hospital).
- Targets and standards for waiting time for appointments with community nursing staff (North West Anglia Community Services Trust).

- Patient satisfaction surveys in nine departments carried out last year and results used to identify areas for service improvement (Essex Rivers Healthcare).

12.9 Benchmarking in R&D

Western companies have long been recognized for their attitudes towards fulfilling the needs of the shareholder first and the customer last. As argued by Motteram (1991):

> On the one hand, a company cannot afford not to develop new products, on the other, to provide such investment would damage cash flow and start to concern investors.

Typically, the key measures used to assess business strength in the West and performance in the marketplace are:

1. **dividend yield**, a short-term return, also providing funding for future investment;
2. **earning per share** comparative to other companies which also indicates trend, and is an indicator of performance against previous years;
3. **price earning ratio** used by analysts for comparisons across companies;
4. **return on capital employed**, a measure of efficiency on how the business is being managed;
5. **current asset turn** – the lower the turn the more risk of obsolescence;
6. **quick ratio** which reflects the organization's ability to meet current debts from current liquid assets;
7. **current assets versus current liabilities** is another ratio for determining liquidity;
8. **capital gearing** measures the potential for borrowing power to fund new projects and new investments;
9. **PV R&D charges** indicate whether sufficient funding is being ploughed back into the business – an indicator of growth;
10. **capital investment level** reflects how well the business is without stretching the resource strategy;
11. **order cover** measures the confidence in future business (Motteram, 1991);
12. **R&D decision** is carried out very subjectively, varies from industry to industry, business to business and fluctuates in amounts from year to year.

12.9.1 The management of R&D

Senior managers are trained to manage all key functions of business organizations such as marketing, human resources, finance, sales, distribution, manufacturing, maintenance, etc. They are not, however, trained at managing R&D activity. Very few textbooks refer to R&D as a complex process that requires proper management. This deficiency in senior managers is recognized by many people. Dr H. A. Schneiderman, Vice President of R&D, Monsanto Company, stated in 1990 that:

> Although most technology dependent companies recognize R&D as a necessary business experience, their executives usually do not understand it as well as they do manufacturing, sales, marketing, finance, human resources, or even legal and regulatory affairs. Yet R&D fuels these companies' earnings growth.

Typically, there are five questions that senior managers have to wrestle with.

1. **How long should it take and how much does it cost?**
   Priorities are often given to incremental innovation with clear, tangible and quick payback.
2. **How to inspire more productive R&D?**
   Senior managers look for winners and successful projects without trying to understand how R&D functions can deploy the right resources for its management.
3. **How to integrate R&D with the rest of the company?**
   R&D is considered very often as a cost centre rather than a profit centre. It is not integrated to functions such as marketing, production and sales.
4. **How to pick winners?**
   Lack of understanding of how R&D should be managed and lack of management means that it is not linked to corporate strategy and corporate goals are not translated in terms of innovation activity.
5. **When and how to terminate failures?**
   Lack of strategic direction, poor resourcing and poor management means that many projects are allowed to drift, absorbing resources and finishing as complete disasters.

12.9.2 Benchmarking R&D for best practice

One of the problems associated with R&D is the level of secrecy surrounding what takes place inside laboratories and R&D departments. R&D is often referred to as 'the black box' containing all the information on future prospects for business organizations. The R&D scoreboard initiative was introduced by the Department of Trade and Industry in the UK in 1991 as part of a strategy encouraging innovation in the UK.

The benchmarking framework developed used the following indicators for establishing comparisons:

- current spend;
- R&D spend per employee;
- R&D as a percentage of sales;

- R&D as a percentage of profit;
- R&D as a percentage of dividends; and
- percentage sales per employee.

The R&D scoreboard indicates various levels of commitment to R&D expenditure but more importantly it determines that there is a strong correlation between R&D expenditure and performance of companies in the marketplace.

12.9.3 Benchmarking R&D management in Japan

Figures published in 1991 from the R&D scoreboard indicate that based on international comparisons using aggregates of the top 100 R&D spenders per country:

- Britain spent £5.9 billion, with £1530 per employee;
- America spent £27.7 billion, with £3730 per employee;
- Japan spent £20 billion; and
- Germany spent £14.7 billion, with £4320 per employee.

Japan, therefore, lies second in R&D expenditure behind the US, demonstrating high commitment for long-term survivability and prosperity. In Japan, it is thought that the competitive threshold from R&D work is further downstream, only at design or prototyping of products and processes. This is not, however, the case of the US where all R&D activities are considered to be proprietary and that competitive sensitivity starts much earlier in the R&D process at the applied research stage (Hellwig, 1991). Japanese companies believe in co-operation and sharing of development work, until prototyping stages. Co-operative projects extend to the involvement of foreign firms as well.

Table 12.8 illustrates the various stages used in new product development. This 'taxonomy of product genesis' (Hellwig, 1991) was used for benchmarking R&D activity between Japan and US.

Table 12.9 illustrates a tentative comparison between the US and Japan (Hellwig, 1991), trying to explain the differences in competitive outlook that often lead to differences in the management of R&D activity. Seven elements were used for establishing the differences between the two countries concerned. The practices found to be different in Japan and which often lead to better R&D management, include:

- R&D activity based on team co-operation and deploying synergies and levels of creativity in the same direction;
- the encouragement of experimentation and creativity – failures are considered as opportunities for improvement;
- the management of innovation is through incremental improvements; optimization of learning, enhancing capability and effective utilization of resources available;

Table 12.8 Taxonomy of product genesis

Category	Subcategory	Objective
Research	Fundamental research	Enhance knowledge/ understanding
	Generic research	Support one or more engineering disciplines
	Applied research	Support a specific engineering area
Development	Product/process conceptualization	Generate product/process options and alternatives
	Product/process design	Select product/process option(s)
	Product/process prototyping	Verify performance/cost objectives
Commercialization	Production/processing	Make (cost-competitively) a quality product
	Testing/acceptance	Verify quality/performance
	Marketing/selling	Generate revenue, profit and market share

- a focus on the process, sustaining capability and consistency to deliver successful products and services;
- focusing on the customer through sustaining high competitive standards, quality, responsiveness and market presence.

Table 12.9 Comparison of business strategies in the US and Japan

	Traditional strategic approaches	
Strategy element	In Japan	In the US
Employee attitude	Team/cooperation	Individual/competition
Risk/reward	Learn from failures	Punish failures
Innovation	Incremental improvements	Breakthrough
Manufacturing focus	Process	Product
Product objective	Quality/utility	Performance/novelty
Product introduction	Sustain market presence	First to market
Business focus	Market share/customer	Bottom line/ shareholder

This brief benchmarking exercise demonstrates that competitive advantages have to be determined through:

- a thorough understanding of business processes, their capability, consistency;
- an understanding of business priorities, desired goals, their communication and translation throughout the business;
- attempts to manage *all* activities with the same degree of focus; attention means a dilution of efforts, effectiveness and a misuse of resources available;
- harnessing creativity through people involvement, teamwork would lead to better R&D output;
- uniqueness, not necessarily associated with the technology/prototype, but how this is applied within business operations;
- success not in inventing, but innovating and achieving total customer satisfaction, increased market share and significant growth.

12.9.4 The drivers of innovation

Traditionally R&D tended to be driven heavily by:

- pure/applied research;
- technologies; and
- defensive and offensive patents.

More increasingly, however, technical success in the protection of prototypes and blue chip technologies is not considered enough. It is the conversion of technical innovation into commerical success that is the sought end result and therefore a good understanding of market conditions and customer requirements is of the utmost importance.

More and more companies are demanding more successful innovations or translation of technical success into commercial success. To reduce the lead time, in the development of technical innovation into successful launches, demands an integrated approach to the management of innovation.

An integrated model for managing best practice R&D would start with a customer-focused strategy and the integration of all key activities (Fig. 12.18).

Best practices in R&D

The following criteria are often found in successful organizations, recognised for their track record of launching successful products and services in the marketplace. These criteria are thought to lead to the successful management of R&D:

- commitment to a long-term marketing perspective;

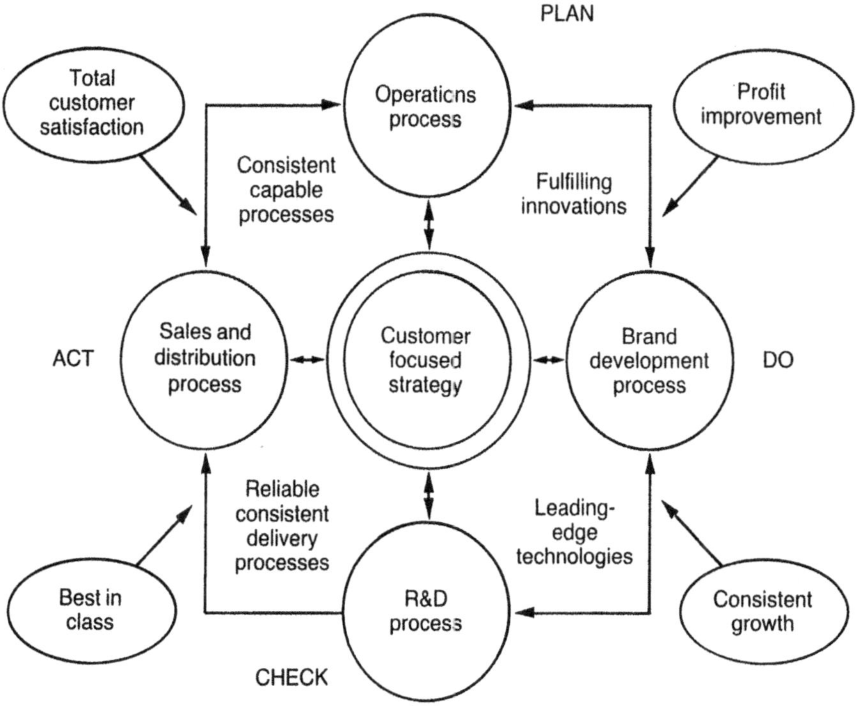

Figure 12.18 An integrated model for managing best practice (R&D).

- using understanding of customer needs as a basis for developing business plans, making decisions for capital investments, for planning operations and developing the required technologies;
- corporate culture that encourages problem-solving the management of opportunities for improvement rather than 'fire fighting';
- having R&D as an integral part of corporate strategy; and
- having strong leadership.

In addition, the following excellent practices were extracted from model companies recognized worldwide for their superior innovative ability:

- having mechanisms to assure translation of R&D results to a sustainable commercial advantage in identified areas;
- nurturing creative, inventive people;
- performing state of the art R&D;
- being knowledgeable in, and responsive to business needs;
- having multifunctional technical transfer teams;
- ensuring that transfer documents fit company strategy;
- early involvement of customers;
- ensuring continuity and developing good personal contacts;
- having realistic goals, timescales; and
- subjecting R&D to quality principles and a never-ending improvement ethos.

12.10 Benchmarking in the environment

There is more pressure on companies nowadays to develop and implement environmental policies and procedures and to start taking the impact of the environmental issues as a critical element of the competitive equation.

Various surveys at corporate level have indicated how important environmental issues are to future business success. A report by McKinsey (Stikker, 1992) was published following a worldwide survey of top executives in 400 companies, asking their views on the environmental challenge. Three key findings were highlighted in the report.

1. More than half of CEOs interviewed admitted that environmental issues are a major business issue for the 1990s.
2. At senior management level, there was considerable doubt as to the feasibility of dealing with environmental issues effectively and yet remain competitive.
3. Over half of the companies analysed were primarily focusing on compliance with the law and trying to prevent incidents from happening.

Another survey was carried out by Booz-Allen & Hamilton in 1991 of 200 senior executives to find out how major companies understand and manage environmental issues (Newman and Breeden, 1992).

- Sixty-seven per cent of executives surveyed recognize environmental issues to be extremely important to their business. This is three times the level reported in an earlier survey conducted in 1989.
- Only 7% felt very comfortable and confident that their companies have comprehensive risk management systems and can deal with major risks.
- Nearly all executives interviewed admitted that they must manage environmental risk better in the future.
- The surveyed companies have also recognized that they must take the lead by moving away from a regulatory-driven management mode to a pro-active mode, strategically driven for establishing competitive advantage.

12.10.1 The environmental issues – what are they?

- **Greenhouse effect**
  In the artificial warming of the atmosphere by man-made emissions, CO_2 accounts for approximately 50% of the greenhouse effect. Globally, fossil fuel combustion leads to 5.7 billion tons of CO_2 emissions per annum.
- **Depletion of the ozone layer**
  The ozone layer is situated 12–30 miles above the earth. It absorbs 99% of UV. CFCs account for 90% of the damage to the ozone layer. CFCs are to be phased out by the year 2000 in 59 countries.

- **Acid rain**
  Acid precipitation damages leaves (which filter acid deposits from the atmosphere) damaging woodland. Acid emissions include SO_x and NO_x gases.
- **Habitat destruction**
  Rain forest constitutes 8% of land area, containing 50% of all known species and anchors climatic conditions. The rain forest is lost at a rate of 200 000 km^2 per annum. It is estimated that 1–50 species per day become extinct.
- **Waste**
  One billion tonnes of waste is produced worldwide each year.
- **Transport**
  Motor vehicles account for 80% of lead emissions and 85% carbon monoxide emissions.

12.10.2 What are the key challenges facing business organizations?

Awareness of major benefits to be derived from focusing on the environment and the pressures of social responsibilities makes environmental issues a corporate debate. There is concern for public image, rising insurance premiums, increased community awareness of the environment, the implementation of stiff penalties and fines for non-compliance with environmental legislation (Polonsky and Zeffare, 1992).

In addition, the emergence and growth in green consumerism, together with changes taking place in consumer behaviour is putting pressure on producers of goods and services to use ecologically friendly processes and ingredients, for the manufacture of ecologically friendly products and services.

There are increasing demands on companies to manage processes, measure performance on environmental issues, generate data and publish information on their performance.

Against this background of developments, it is very clear that increasingly the environment is going to become the major discussion point in boardrooms. Certainly, management of this issue both strategically and operationally would create competitive advantages.

A new standard for managing the environment has been developed (BS7750) and many organizations have taken a keen interest in it. Environmental auditing is a critically important activity and this aspect of performance measurement is often used for drawing together action plans and for the planning and development of effective strategies.

Booz-Allen benchmarking study

This study was conducted in 1992 (Newman and Breeden, 1992) to gain some understanding of how environmental management relates to corporate strategy. Eight large companies were targeted for this study, for their excellence and leadership on environmental management as recognized by various bodies such as the Council on Environmental Quality, the World Environment Centre and the Global Environmental Management Initiative (GEMI).

The study was also aimed at capturing information on practices these leading companies use in ensuring successful management of environmental issues. The sample included AT&T, Chevron, *The Los Angeles Times*, McDonalds, Pacific Gas & Electric Company, 3M, Rohm and Haas, and IBM. Amongst the top key findings are the following.

- Pacific Gas & Electric (with plants) focuses on air emissions.
- McDonalds (without plants) focuses on solid waste, particularly packaging.
- 3M focuses on new product development and the environment, for sustainable development and future regulations that might affect their product lines.
- Rohm and Haas, in the manufacture of polymers, is concerned by air, water and solid waste pollution and is amongst the top leaders in preventing groundwater contamination.
- *The Los Angeles Times* is benefiting from using more than 80% recycled newsprints and has saved a lot of money by cutting down on emissions through switching its delivery fleet to propane vehicles.
- AT&T saves money through promoting alternative materials to CFCs used widely in electronics. AT&T has achieved 100% elimination of CFCs at six of its plants.

Other examples of best practice in environment management

Procter & Gamble was the first company to develop refillable pouches for fabric softener in response to the public's concern over solid waste. They met consumer demands by taking this initiative but also reduced product packaging. Procter & Gamble has also continued innovating in this area through product combinations (detergent with bleach or fabric softener) using completely recycled plastic containers and recyclable packaging (Greeno and Robinson, 1992).

Black & Decker takes back a product when its life runs out by assuming the responsibility to recycle the product (and the battery) for the customer (Greeno and Robinson, 1992).

Henkel launched phosphate free detergents in 1986 and this pioneering work led to detergents containing phosphates to be phased out in Germany in 1989. Other detergent producers followed (Barnett, 1992).

Volkswagen are pioneering the concept of an expanded product life cycles programme to go beyond sales and service and towards total remanufacturing. With its 3V policy (vermeiden, veringern and verworten) meaning, prevention (reducing solvent emissions by switching to water-based paints), reduction and recycling (a target of 100%), Volkswagen is showing best practice in this field (Simon, 1992).

12.10.3 Environmental leadership role of transnational corporations

Many transnational companies are implementing environmental management programmes similar to those introduced in the West. **AT&T** seconded a senior environmental manager to China, Hungary and Russia to advise their governments on CFC alternatives. **Chevron** is leading a petrochemical initiative to share environmental know-how with Nigeria (Westcott II, 1992). **Northern Telecom** is helping the Mexican government with technical and managerial know-how regarding non-CFC cleaning processes in electronics manufacturing. **Union Carbide** is helping Thailand in hazardous waste treatment. **3M** is training managers from Czechoslovakia, Hungary, Poland and Turkey on environmental management.

12.10.4 Managing the environment for excellence – an integrated approach

In 1991, The Chartered Association of Certified Accountants (CACA) established the Environmental Reporting Awards Scheme (ERAS) to identify best practice in environmental management and to encourage development, implementation and the sharing of learning amongst industrial companies. The winners of 1991 included **British Airways** and **Norsk-Hydro** who were found to excel in the following ways (Gray and Owen, 1992):

- a systematic and thorough review of core business activities;
- a balanced view of environmental performance;
- independent attestation of the environmental report;
- widespread distribution of the report; and
- a commitment to carry on publishing further environmental reports in the future.

The winners of the 1992 ERAS Award are **British Telecom (BT),** the UK's principle supplier of telecommunications services, handling an average of 90 million telephone calls a day. During their financial year ending March 1992, BT had a turnover of £13 337m, employed 210 500 people, invested £1 895m on the installation of digital exchanges and transmission technologies and spent £240m on R&D.

The environmental management system illustrated in Fig. 12.19 reflects how BT management is committed to environmental achievements. The environment is at the heart of policy/strategy development, as reported by BT's Deputy Chairman, Mr Mike Bett:

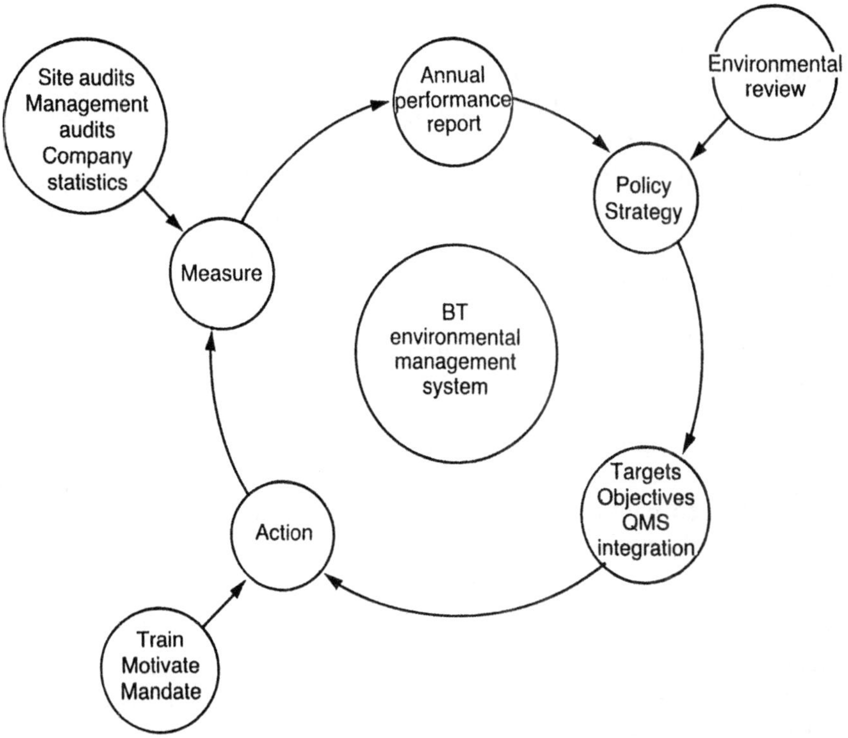

Figure 12.19 A leader's environmental management system: the case of BT.

> Our overall target for the future is for environmental considerations to play an increasingly important part in the company's planning and management process.

The model developed by BT is based on the new standard BS7750. BT is working towards BS7750 recognition. It is committed to integrate its environmental management system (EMS) into its quality management system. The key target for BT is to have all main functional units carry out environmental assessments and audits on all their activities.

12.11 Benchmarking quality systems

12.11.1 The role of quality systems

Quality systems such as BS5750 and its equivalents (ISO9000 series and EN29000) play a significant role in securing markets for organizations by building in capability to deliver quality products and services, creating consistency and achieving goal congruence through the establishment of customer–supplier chains.

Processes are operated and managed through series of procedures. The mechanism for operating quality systems is to write down what is done, do what you say is carried out, and record what is done. Organizations can choose which parts of quality systems such as BS5750 they apply for. Part I includes 20 operating procedures with the inclusion of design and servicing. Part II has 18 operating procedures excluding design and services.

Quality systems bring a wide variety of benefits including improved utilization of materials, waste reduction, speed reduction, delivery reliability increases, increases in efficiency, supplier partnerships and customer satisfaction.

Management systems only become dynamic and lead to quality improvements through the use of quality audits and management reviews. Quality audits are a requirement of the system. They involve an ongoing review of the system at suitable intervals by management and appropriate corrective and remedial action taken when deficiencies are identified.

12.11.2 Benchmarking quality systems: the PERA International survey

A survey was conducted by PERA International (based in the UK) in 1992 to assess implementation and benefits of BS5750 (Booth, 1993). A series of telephone interviews based around a sample of 2317 organizations was undertaken. It was found that, on average, it takes 18 months to achieve registration. As a result of registration, 40 000 firms will achieve improvements in operational efficiency and 30 000 manage to improve their marketing. Others believe that by having BS5750 they have managed to increase their profitability.

The PERA International Survey concluded that 89% of surveyed companies believed that the introduction of a formal management system has had a positive effect on operating efficiency, 48% of companies claimed improved profitability, 76% claimed improved marketing whilst 26% of firms claimed improved expert sales.

12.11.3 Benchmarking quality audits

There are two requirements of quality systems:

1. **internal audits** to ensure that the system and the standards conform and that the practices in place comply with the documented procedures; and
2. **external audits** by an accredited assessor. This is in order to add value to the audited company rather than simply using the audit as a control tool.

Companies that achieve registration use the audits as a means for not only non-conformance detection but also as tools to carry out continuous improvement in all the processes and functions.

A benchmarking study was recently conducted at Bradford University in the UK (Barthelemy, 1993) to assess the use of internal audits in a small number of leading companies with the view of determining best practice. The project was also intended to analyse the interface between internal and external audits and recommend best practice. Five companies were used in this project: Avis Rent A Car, Elida Gibbs Ltd., IBM, ICL and Kodak UK Ltd. The study found that internal audits were used for non-conformance purposes rather than continuous improvement. The non-conformance orientation is due to the fact that the types of audits performed are functional and follow the standards, rather than trail products and processes (Barthelemy, 1993). There is a lack of leadership and non-conformance audits are an easy option. Although auditors were found to enjoy the same training, it appears that in some sites people are process-oriented and others seem to be much more concerned with non-conformances and deviation from the standards. Variation in the efficiency of audits was found to occur between sites. Perhaps internal benchmarking would be extremely useful in helping the sharing of best practice and also in optimizing and standardizing methods and results. Audits tend to be used for rectifying short-term problems without necessarily trying to implement prevention programmes with long-term solutions.

The benchmarking study found that external audits tended to have many shortcomings such as lack of preparation for the audits, a high degree of variation in the quality of the auditors, 'tick in the box' audits, and no follow-up of action or feedback.

External auditors do not communicate effectively with the audited companies. A large proportion of the time is spent examining the reports from previous audits. Auditors worsen the situation by changing almost every audit, hence increasing the degree of variability of the output. There is an information gap and there has to be a way of closing it.

Table 12.10 illustrates the benchmarking summary of ISO9000 audits in the companies examined.

12.12 Benchmarking in the public sector

Various areas of the public sector use the art of benchmarking – without perhaps knowing that they do so and without necessarily exploiting the full potential from benchmarking exercises – in order to strengthen processes and continuously introducing best practice. Local authorities (governments), for instance, compile trends of performance and compare them against national average. Good results – or disappointing ones – are then explained by referring back to the processes concerned. In more recent years, local authorities started to incorporate European trends as a result of the common market amongst European countries.

One example in this sector is the Highways Services. Each local authority is committed to reducing highways and transportation accidents through investment and various programmes geared towards minimizing

Table 12.10 Benchmarking of ISO9000 audits

	A	B	C	D	E
% people trained for auditing	3	2	(a)	10%	3.75%
Number of internal audit/dept/year	1	1	1	1	1
Purpose of audit (b)	NC&CI	NC&CI	NC	NC&CI	NC&CI
Result of audit (b)	NC	NC(&CI)	NC	NC(&CI) depends on site	NC&CI
Type of audit (c)	IS&ID	IS&ID	IS&ID	ID	IS&ID
Audit organisation responsibility	auditor	auditor	auditor	auditor	auditor
Documents produced	Report CAR	Report CAR	Report CAR	Report CAR	Report CAR
Follow up made	Site Manager	Quality Manager	Site Manager	Site Manager	Div. Manager
Level of top management involvement	medium	medium	high	high	medium
% Non conformances corrected as agreed	(d)	26–73% (e)	(d)	95%	65–70% (f)
External assessor	Veritas	BSI	BSI	BSI	BSI
Number of external audits/year	4	2	2	2	2
Number of times each dept audited	2 (g)	1 (h)	1 (h)	1 (h)	1 (h)
Satisfied with external assessor	Yes	Situation improved	No	No	Situation improved

(a) This information is considered sensitive by C. Its publication has not been authorized.
(b) NC: Non conformance
CI: Continuous improvement.
(c) IS: Inter site
ID: Inter department.
(d) No figures available.
(e) The results are fairly irregular varying over time from 26 to 73%.
(f) The measure is: 'corrected first time' and not 'corrected at the agreed date' because the follow up audit may not take place until some time after the corrective action completion date.
(g) At A, every department is audited twice a year, internally and externally.
(h) In the other companies, each department or process will generally be audited once a year. It can be audited a second time, externally by BSI (British Standards Institute).

or eradicating accidents of any kind. As expressed by the following statement from one of the local authorities in the UK:

> The council will do everything in its power to reduce accidents, but road users are ultimately responsible for the greatest reduction by adjusting their driving techniques to suit the road conditions.

The benchmark set by the UK government is to reduce accidents by one third by the year 2000. On average it is found that each accident produces 1.3 casualties. In order to achieve the target set by the government, schemes were set up aimed at educating and changing drivers' attitudes through better awareness, provision of information and appreciation of hazardous situations. Appropriate legislation was also introduced. For instance, seat belt legislation was found to have made a contribution in reducing the number of casualties from accidents. In addition, stricter driving standards for bus and coach drivers are leading to lower casualty records.

There are also economic implications for maintaining accident rates at a low level. It is thought for instance that the average cost per accident is around £26 000. In one of the district communities, the cost of accidents in 1992 amounted to £37.5 million.

Figure 12.20 illustrates casualty trends for the period 1985–92, benchmarking performance at district, regional and national levels.

Figure 12.21 illustrates injury trends for the period 1985–92 at district, region and national levels.

Causes of accidents are found to be directly attributable to human error 95% of the time. A smaller number of accidents are found to result from factors such as environmental or vehicle defect factors.

Table 12.11 illustrates a comparison of casualties by group. The statistics show a five-year average, the trend, and the percentage change from the previous year.

Figure 12.22 illustrates casualty trends over a five-year period between various groups.

12.13 Benchmarking in a multinational/monopoly situation

Benchmarking in large organizations such as multinationals or companies that are in a monopoly situation is highly applicable. Most large organizations are either retracting or becoming global and as such they are applying benchmarking very aggressively to achieve their goals.

One major sector where benchmarking is being applied significantly is in telecommunications. This sector is growing despite the world recession; the volume of telephone traffic is still increasing and most operators are aspiring to become global. This is spurred on by the fact that governments are relaxing the rules and privatizing their state-run telephone companies to encourage competition. In addition, the exploitation of new technology

Table 12.11 A comparison of casualty per group – benchmarking performance over 5 year span

	1985	1986	1987	1988	1989	1990	1991	1992	Five year ave. 85–9	Percentage change on 5 year average	Percentage change on 1991
	444	436	429	442	426	481	422	405	435	Down 7%	Down 4%
	84	98	97	97	126	100	113	118	100	Up 18%	Up 4%
 Incl. pillion	345	378	279	264	244	238	165	151	302	Down 50%	Down 9%
 Driver	483	435	490	529	585	617	617	630	504	Up 25%	Up 2%
 Passenger	424	339	383	388	432	423	380	423	393	Up 7%	Up 11%
Others	146	121	166	141	118	131	123	152	138	Up 10%	Up 23%

	1985	1986	1987	1988	1989	1990	1991	1992
Local ——	100	97	98	98	105	106	97	98
Regional ——	100	102	102	107	113	112	106	106
UK -----	100	101	97	101	106	105	96	

Year

Figure 12.20 Casualty trends – benchmarking performance: local, regional and national trends (1985–92).

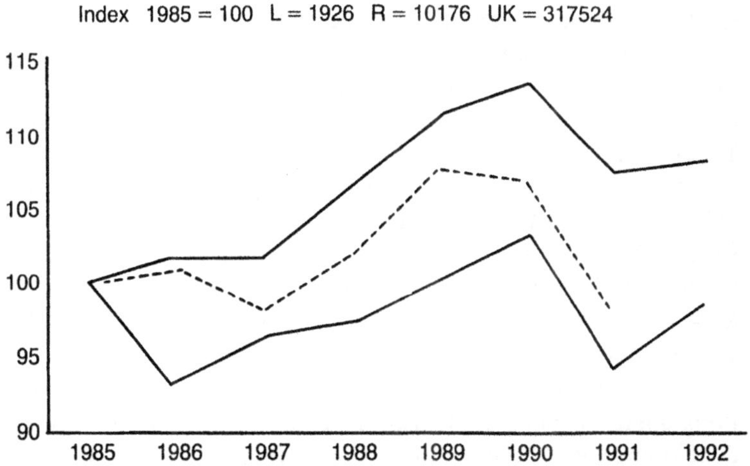

	1985	1986	1987	1988	1989	1990	1991	1992
Local ——	100	93	96	97	100	103	94	98
Regional ——	100	102	102	107	112	114	108	109
UK -----	100	101	98	102	108	107	98	

Year

Figure 12.21 Injury trends – benchmarking performance: local, regional and national trends (1985–92).

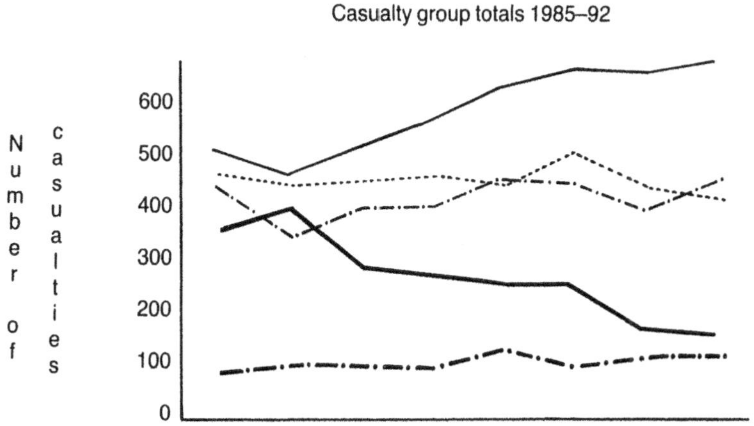

		1985	1986	1987	1988	1989	1990	1991	1992
Cyclists	—·—·—	84	98	97	97	126	100	113	118
Mtr cycle/pillions	▬▬▬	345	378	279	264	242	238	165	151
Pedestrians	–·–·–	444	426	429	442	426	481	422	405
Car passengers	....	424	339	383	388	432	423	380	423
Car drivers	———	483	435	490	529	585	617	617	630

Year

Figure 12.22 **Casualty trends – benchmarking performance per group (1985–92).**

means that costs can be reduced drastically, efficiency improves and the quality of services becomes much better.

Advanced technologies in telecommunications include the introduction of fibre-optic cable and digital exchanges, the use of intelligence networks combined with advances in radio and semi-conductor technology (Dixon, 1992). Competitiveness in telecommunications is expected to rise as more and more privatization is going to take place. According to Booz-Allen Hamilton Consultants, between 1991 and 1993, telecom companies in a further 26 countries, responsible for 95 million lines, are expected to be privatized (Taylor, 1992). The newly formed companies will start to look at global customers through acquisitions, strategic alliances or as stakeholders.

Strategic benchmarking is used aggressively in the large telecommunications operators such as British Telecom and AT&T. BT, for example, are developing their 'cyclone' programme which is a network intended to help international corporate customers by offering them text, data and video services as part of a 'virtual' private network. These virtual private networks through the use of special software, common transmission and switching facilities can be shared between different customers. BT intends to capture a big slice of the growing market of 'outsourcing' where they will be expected to manage private networks of international corporations that have subsidiaries and overseas offices all over the world. Outsourcing is thought to bring savings of up to 20% off the corporate communications bill (Williamson, 1992). Large companies are increasingly outsourcing telecommunications networks because this is not considered as a core activity of their business. AT&T, on the other hand, who invented cellular technology and are a leading supplier of cellular equipment, are developing strategic alliances to dominate globally in long-distance and international markets and also in equipment manufacturing (Dickson, 1992).

Benchmarking in the chemical industry

Unlike the telecommunications sector, the chemcial industry is retracting. Recently, for instance, ICI was reorganized to create a new subsidiary (Zeneca) to handle its drugs and fancy chemicals, selling off in the process its petrochemicals and plastics, to leave it concentrating on specialized chemicals such as paints and acrylics. Union Carbide, on the other hand, one of America's biggest firms has shrunk to less than 50%, whilst Monsanto, another large company, has got out of commodity chemicals altogether. The chairman of ICI in relation to commodity chemicals was reported in *The Economist* to have said: 'The chemical world has gone to hell in a handcart.'

Table 12.12 illustrates a list of the world's biggest chemical companies, at 1991 sales figures.

BP Chemicals is a wholly owned subsidiary of British Petroleum PLC. It has the following vision:

> Through the skills of our people and the quality of our science and technology, we will become the world leaders in our chosen sectors of the chemical industry.

BP Chemicals strategy covers the following areas (Priestley, 1993):

- strength in business portfolio;
- technology and marketing to serve customers' needs;
- globalization – Asia Pacific Region first;
- industry leadership in health, safety and the environment;
- releasing the full potential of staff as part of total quality; and
- operational priority – to improve performance.

BP Chemicals initiated a benchmarking process particularly to improve its performance and harness the potential of people and optimize the utilization of its resources so that all strategic goals could gradually be achieved. The purpose of benchmarking at BP Chemicals is (Priestley, 1993):

- to achieve best-in-class cost structure;
- to ensure cost competitiveness to attract investment;

- to provide competitive advantage for each product
  - customer satisfaction, and
  - quality, reliability, delivery and cost; and
- to seek continuous improvement.

BP Chemicals has benchmarked in the following areas:

- organigrams and manpower activity;
- organizational structure and philosophy, e.g. quality application, management layers and devolved accountability, career structure and retention of experience;
- purchasing, e.g. supplier evaluation and selection;
- operations; and
- maintenance (maintenance approach – predictive, preventative or breakdown).

BP Chemicals has used benchmarking both internally between its various sites and externally both within the chemical industry sector and outside the chemical industry, seeking best practice in total business process management.

Table 12.12 World's biggest chemical companies (1991)

	Sales, $bn
1 – Hoechst	31.1
2 – BASF*	30.8
3 – Bayer	28.0
4 – ICI	23.3
5 – Dow Chemical	18.8
6 – Rhone-Poulenc	18.1
7 – Du Pont*	17.9
8 – Ciba-Geigy	15.5
9 – Elf Aquitaine	14.0
10 – EniChem	11.7
11 – Shell*	11.2
12 – Sandoz	9.9
13 – Akzo	9.9
14 – Exxon*	9.2
15 – Mitsubishi Kasei	9.2
16 – Monsanto	8.9
17 – Sumitomo Chemical	8.7
18 – Solvay	8.1
19 – Huls	6.8
20 – BP*	5.5

*Oil assets excluded

Benchmarking in the airline industry

In this industry competition has intensified over the past few years. Small carriers are finding it extremely hard to compete and large carriers are seeking to globalize their operations and dominate worldwide.

British Airways have extensively used the art of benchmarking to improve their various processes, particularly those which impact most on customer satisfaction. One such process which was benchmarked is the passenger arrivals process. Weaknesses were identified in the arrivals process (Davies, 1992):

- no special requirements pre-warning;
- limited staff availability immediately;
- absence of clocks;
- directions only in English;
- unattended immigration queues;
- poor ramp directions to belts;
- baggage belt congestion;
- inconsistent baggage attendance;
- guidance to lost bag counter; and
- uncoordinated baggage announcements.

British Airways broke their arrival process into the following key elements:

- disembarkation,
- transfer,
- immigration,
- baggage,
- customs, and
- onward.

Seven benchmarking competitors were selected and cross-functional visits conducted. The seven competitors were found to have strengths in the following areas:

- arrivals utilization;
- baggage reclaim;
- use of multifunctional staff;
- lost bag facility;
- multilingual signage;
- reservations/enquiries;
- immigration queue-combing; and
- off-pier steps/lounges.

As a result of their benchmarking exercise, British Airways managed to put together an action plan which led to the following results:

- procedure changes negotiated;
- new elements of the service and the style used were changed;
- specification of service was enhanced; and
- better performance measures put in place.

Overall, British Airways understands the arrivals process much better, it is operated as an integral process with cross-functional participation, using best in class methodologies and a focus on process improvement and customer satisfaction.

13

Critical factors in benchmarking

Benchmarking is not about methodological issues only; it is more about how the methodologies available are implemented and applied to particular situations. Where do you gather information from? Whom should you use for the benchmarking exercises themselves? What partners should you look at in the process of selecting the projects? Who should champion the whole introduction of benchmarking? How to make it sustainable and thus a powerful tool in changing the culture of the organization? These are the critical factors which facilitate the introduction of benchmarking. This chapter discusses them in more detail.

13.1 Organizational factors

Organizations are competing in different market situations. Some are lagging behind their competitors and would like to see the gap closed. Other organizations are sitting very comfortably at the top of the league and dictating how competition is going to take place. The various problems to be looked at in terms of benchmaking will therefore differ from one organization to another. Benchmarking in many situations needs to be applied urgently since many organizations are seeing their market share eroded, competitors outbidding them and outperforming them in the marketplace, and their efficiency and effectiveness not up to the desired standard. For these organizations, benchmarking needs to be introduced urgently and therefore would require a lot of resources, a lot of commitment and company wide contributions to make it successful. Other companies, however, are in a very comfortable situation and are therefore able to introduce benchmarking and learn from it without needing to achieve results so desperately. Whichever position they find themselves in, companies have to realize that competition does not slow down or stop for them. Benchmarking has to be applied now – tomorrow may be too late. The business climate will never be right.

Benchmarking is not a package that can be bought off the shelf. Companies need to have some knowledge and understanding of how benchmarking can be applied and the implications when introducing it for the first time in their business operations. Benchmarking is a very powerful agent for change and once introduced it has to be made sustainable. This area is critical since most organizations will have been exposed to benchmarking for the first time, and may require external assistance to increase the pool of knowledge and the know-how to introduce benchmarking. As with total quality management, benchmarking has to be positively managed otherwise it will be treated as industrial tourism. Action has to be the ultimate outcome from all the benchmarking exercises, and the exercises will have to be reviewed on a regular basis using the continuous improvement route of plan, do, check and act.

Benchmarking is a corporate exercise. It is very important to avoid the back door approach, malpractices and industrial espionage. Benchmarking, for it to be sustainable, is not an exercise driven by individuals. The sharing of information has got to be carried out in an open manner with the support and encouragement and awareness of top management. A corporate organization must be willing to share that kind of information, rather than have pieces of paper being passed from one organization to another in total secrecy. The authors have come across many situations where advice given to organizations is: 'If you cannot gain access at the top, try at a lower level.' The fact of the matter is that if top management is not willing to share or is unable to appreciate the importance of benchmarking and its usefulness, this clearly demonstrates that these organizations are reactive and perhaps have nothing to share but a lot to hide.

Is sharing information the most critical issue? Understanding where the competitive advantage is and where the weaknesses are, will enable organizations to use benchmarking to protect those weaknesses and to reinforce the competitive advantage. The Japanese adopt this principle very comfortably and confidently; they do not necessarily see that by sharing information they are eroding their competitive advantage. The most critical thing for organizations, therefore, is to determine from the onset where the competitive advantage is, what the distinguishing features of their products and services are, and where they excel in comparison with the rest of the league. In addition it is very important for organizations to gather the voice of the process, and through the voice of the customer determine weaknesses in areas where resources need to be devoted and where the knowledge base needs to be deployed in order to close the gap.

13.2 Benchmarking and TQM: a positive link

The chicken and egg situation: what should come first – TQM or benchmarking? Many organizations will admit that they have not really embarked on the introduction of TQM and because of this they say that benchmarking is not a practice for them. However, there is a great inter-dependency between TQM principles and benchmarking; in fact they are an integral part of one another, and as such one could trigger the birth of the other. Having TQM in place will enable organizations to use the art of benchmarking. There is no reason why starting with benchmarking should not trigger the introduction of TQM on a wider basis. For example, National Power (an organization which was formed alongside Powergen as a result of the splitting up of the CEGB in the UK), through the use of focus groups who are small groups of enthusiastic individuals, has found that benchmarking works. This form of benchmarking offers quick returns and can be linked very easily with other change techniques and does not require a lot of resources. Focus groups concentrate mainly in a broad plant or functional area and are staffed by people who understand the process and therefore are going to make change happen. Focus groups have helped the organization move away from comparing metrics to understanding and comparing processes. In view of the success achieved through the use of focus groups, National Power intends to roll out the idea of focus groups throughout the whole organization, and to broaden the range of projects that benchmarking would be looking at. They plan to start to introduce the idea of continuous improvement philosophy as a way of conducting business.

Benchmarking is not a process where success is achieved in one leap. It is about incremental change and the gradual process of moving from continuous improvement to continuous learning. There are two aspects to benchmarking. One is positive quality where the exercises would determine that there is a positive gap between organizations and their competitors, in this way benchmarking will enable organizations' design strategies to maintain the advantage to sustain that level of superior performance. The second aspect to it is negative quality, that the benchmarking gap is not in favour of the organizations themselves but their competitors. Using total quality management principles and problem solving techniques the gap can be closed gradually to put the organizations at the same level as their competitors.

13.3 Partner issues

This is one of the most critical areas of applying benchmarking. Organizations need to determine the extent of access, the extent of information availability, the extent of sharing of resources, the knowledge available within partner organizations and so on. It is often assumed by organizations that what is required from them is available within the partner organizations and also that the partner organizations are willing to share openly all the requested information. In addition it may be assumed that all partner organizations may have all the knowledge base required for the benchmarking exercise. In reality, however, this is not the case. This is why it is often desirable to look at more than one partner for doing benchmarking exercises. The planning stages of benchmarking exercises are therefore crucial and a common data collection strategy agreed by both partners will always lead to successful outcomes. The choice of the first benchmarking project with partners is critical. It has to lead to a WIN/WIN situation so that other more ambitious projects can be tackled with the same level of enthusiasm, commitment and resourcing by all the partners concerned.

Benchmarking exercises have to be focused on the collection of useful information rather than masses of data. This is because data would lead to increased frustration, misinterpretation, misuse and abuse and complexity of comparisons will increase. However, more refined information will lead to an increased ability to make accurate decisions and bring about the necessary change. Comparabilty would become much clearer and the exercise much more useful. Organizations are always willing to share data, but the teams involved in any benchmarking exercise must sift the masses of data to produce the refined information which is going to lead to decision making and bringing about change to the processes examined. The most critical thing to do is ensure individuals will not become swamped with information and that the process is one of avoiding 'nice to know' and stick to 'need to know'.

13.4 Resource issues

The question of resourcing benchmarking exercises will always be a key one. However, commitment to total quality management and continuous improvement should ensure that organizations will earn the required resources for conducting benchmarking exercises. First they need to establish their *entitlement*. This means that organizations need to establish internal standards of performance and identify weaknesses in the process which do not allow them to reach optimum levels of performance and thus set about closing the gap. Secondly, they need to *close the gap* through earned resource. Once process visibility and progress visibility become established, organizations will be in a position to determine the level of resource required for benchmarking exercises and because of great improvements achieved in terms of cost reductions, the resource earned can be used to close the gap externally by achieving high standards of competitiveness. Once organizations have achieved high standards of competitiveness, extra benefits are accrued and some of those can be deployed towards resourcing more ambitious benchmarking exercises. The key point to remember here is that entitlement resources could represent as much as 50–60% and most organizations cannot therefore make the excuse that they have no resources available to carry out benchmarking exercises.

Most organizations use industrial visits as a means of learning and as a means of capturing practices which can be useful to their operations. Industrial visits could, however, become very costly and would require vast amounts of resources, and this could be another excuse for smaller organizations that have limited resources to carry out benchmarking exercises. A very useful alternative, however, is what is referred to as surrogate industrial visits. This is the use of video tapes of successful organizations describing the benefits and the management of their various processes. An example of this application is the Post Office Counters Limited in the UK who have managed to find a partner in the USA willing to share with them a video they had prepared for their internal utilization describing how they manage and improve specific processes. Post Office Counters Limited have used the video and the process described in the video is a real one, and they managed this project as a normal change project. This enabled them to design a new process through benchmarking their process weaknesses which they have identified to be specifically in the areas of recruitment and training. The new process was then tried out in ten offices throughout the UK and various performance measurement indicators have been used to reflect on the changes taking place. As a result of these changes, more developed specialized training was given to all the employees concerned. The video proved to be a very useful tool for internal communication as well. Amongst the various benefits from this exercise are:

- 80% decrease in window related customer complaints;
- 53% decrease in complaints about long lines;
- 62% increase in compliments;
- 67% felt it made the line move faster; and
- 25% liked being greeted and receiving assistance in the lobby.

This example illustrates the point that industrial visits, whilst desirable, do not have to be undertaken; there are always alternatives which can be used with the same degree of success and which deliver the same quality of changes sought and needed by organizations.

13.5 Communication/planning issues

When introducing total quality management many organizations devote time to communicating the meaning of the philosophy to all their employees. The implications of using it and not using it for their business operations is explained with the objective that everyone becomes committed and everyone buys in to the introduction of this change process. Unfortunately the same degree of interest is not often given to the introduction of benchmarking. Very few organizations describe how, for example, they raise the level of awareness of their employees of the need to conduct benchmarking and the implications of using it and not using it. It is therefore essential that a similar strategy is developed for communicating the benchmarking message and getting everyone to understand that working internally can only give incremental changes and incremental benefits through increasing effectiveness. The only way to sustain competitive advantage and achieve high and superior performance levels is through external comparisons.

In order to raise the necessary degree of awareness, it is essential first of all to pilot the use of benchmarking in a specific area. It is very important also to train people on methodology, data collection aspects, partner aspects and reporting aspects amongst others. This has to be followed by launching the practice of benchmarking organization-wide once everybody understands it and has been trained in its use as a technique. As far as senior management is concerned, the right planning and review processes must be in place to ensure that benchmarking is gradually integrated, as an essential element of managing in the 1990s, alongside the total quality management philosophy.

Benchmarking, to be effective, has to be deployed strategically and is therefore not something that can be bought off the shelf. It has to reflect an organization's fears, aspirations, the vision mission and critical success factors and so on. The process of decision making has to be re-examined if benchmarking is to lead to action plans and efforts which are going to close gaps in competitive terms.

There are different ways of benchmarking in the UK, Japan and the US. The business community operates in different ways from one country to another. The openness in sharing, the trust, the amount of co-operation, the willingness to exchange information, will vary from one country to another. In the UK for example, the pace of introducing benchmarking is very slow, and reflects the existing culture of doing business. However, as more and more organizations are becoming alert to the need for benchmarking and its importance to them, there is more and more willingness to open doors and share, perhaps only on a limited basis. Nevertheless this signifies a new change in the culture of managing businesses in the UK.

13.6 Customer issues

As with internal process management, benchmarking has to be focused on the customer and benefits for the end customer. Benchmarking is an extension of the standards of effectiveness and the search has to be focused externally all the time. It is about closing competitive gaps and nothing else. It is also about innovation, it is about bringing new practices and changing process behaviour and organizational behaviour so that the techniques used, the methods adopted, the way of managing reflects state of the art thinking and best practice. Benchmarking is not just about giving, but also about taking; and as such it is more beneficial to learn from good companies, pioneering companies, best in class companies. As described earlier, benchmarking is about protecting competitive advantage, but also it is about protecting weak processes by strengthening them through new learning, new practices, new methods, new techniques.

Total quality management is essentially about people productivity. Benchmarking, because it is essentially an integral part of total quality management, helps people enhance their knowledge pool.

The establishment of competitive parameters is aimed at tomorrow and not today, and as such the practices that the benchmarking exercises establish should reflect state of the art thinking. The challenge for benchmarking is for it to remain sustainable. To do this it has to have changed the organization's way of thinking and way of operating so that people are operating with the customer in mind and understanding that the market changes very quickly. Benchmarking is a stimulant to creative thinking and also it is very useful for measuring ideas and their worthiness. Encouraging people to benchmark externally will enhance employees' creative contributions and therefore lead to great benefits for their organizations.

Benchmarking is essentially about achieving superior performance in the long term. In the short term, however, it may be about frantic catching up and trying to close negative gaps between organizations and their competitors. However, that is not the end of benchmarking; frantic catching up is not going to lead to superiority. Organizations have to understand that, once the gaps are closed, they have to be in a position to leapfrog competition and sustain superiority for a long term. This is the real challenge for organizations and this is the ultimate measure of culture change. The choice of the process and the choice of the project has to reflect a specific need and also the position of the company at the time, taking into account market conditions, internal conditions, resource issues and strategic issues amongst others.

13.7 Implementation issues

Benchmarking is not a restrictive exercise. On the contrary, it encourages people to use their creativity and innovativeness to the optimum level by questioning everything that they do, and the standard of the outputs from their contributions in line with strategic direction. As such, benchmarking creates flexibility and paves the way for continuous learning. The people who are sent on benchmarking exercises have to understand the implications of the decisions they make. They have to have integrity, but also they have to have the power to make critical decisions with full understanding of the implications for their organizations.

Benchmarkng is reciprocal and is about a WIN/WIN situation with the partners. Data and information exchange is essential. Any review of benchmarking exercises should reflect that the exchange has led to mutual benefit, and that there is positive change within the various partner organizations.

In the past, senior management has tended to have an obsession with quantitative information and most decision making and thinking tended to be based on the analysis of quantitative data. However, benchmarking could generate quantitative or qualitative information. The information is the outcome of the exercise. One should not, therefore, design the benchmarking exercise in the first place to only generate quantitative information. This is very restrictive and may overlook the opportunity of *learning* which is an essential outcome for sustaining strength and competitive advantage.

Who should do the benchmarking exercise? The teams selected for doing the benchmarking exercise have to be enthusiastic individuals, have to have communication skills, a thorough understanding of their business lines, the direction of their businesses and the strategies concerned. They must also have a thorough understanding of their markets and the changes in their markets, and more importantly, a very good understanding of the processes concerned, both technically and in management terms. They also have to be individuals who are committed to the change and who will produce an action plan; a strategy for closing gaps and changing areas of the processes where best practice has been captured. Having a thorough understanding of the process route, how it is operated, controlled, managed, its outcome and the link with other processes and with the strategic thinking will help to decide on the design of the questionnaire and the right questions to ask.

In some instances, the teams concerned could find the visits to partner sites very disappointing in relation to the specifc processes that they wanted to examine in the first place. Rather than terminate the process of exchange too hastily, the team should be encouraged to look beyond the brief. That is, look at other areas where there is an opportunity to learn, and an opportunity to capture good practices.

13.8 Futuristic issues

Processes do not have to be comparable in the way they look, but, more importantly, in the way they deliver. One could therefore compare a mechanical process against a fully automated process against a manual process. The common thread between the three different processes is the output and how the various components are put together to deliver the output.

In the past, processes tended to be designed, implemented and managed with a long-term view. In the West, for example, the lifetime given to a mechanical process would be approximately 10–15 years. This kind of practice tended to discourage people from looking elsewhere, looking outside, bringing in new learning, because the decision has been made to use the equipment for a long term. In the 1990s, however, benchmarking has to be an integral part of process management and operators and managers alike must be encouraged to look outside for new learning, to bring about the necessary changes. The focus must always remain on the customer first, and second on the output of individual processes to achieve optimum levels.

It is critical to know what the benchmark for the processes concerned is. What are the standards sought and the standards achieved? Having those two measures will enable organizations to work towards the benchmark once they have learned new practices from their partners.

Developed relationships with partners have to be nurtured and the contact kept alive. An interest has to be maintained between the two corporate organizations so that future opportunities can be explored and exchanges can be carried out on an ongoing basis. It is painstakingly difficult to generate relationships in the first place. It is therefore wise for organizations to keep in contact with their partners on a regular basis.

On more ambitious projects which are very interesting and essential but very demanding in terms of resources an agreement could take place between the various partners concerned to conduct the projects on a modular basis, reviewing benefits and resource consumptions before deciding on the next stage. Starting ambitious projects could lead to disastrous results and could cost the organizations concerned a large amount of resources. Managing complex benchmarking projects has got to be carried out in the appropriate manner.

Total quality management is about performance measurement and visible progress. The same applies to benchmarking, it is about progress visibility. If there are situations where the result of the benchmarking exercise has led to no change, this needs to be justified thoroughly. If, for example, the teams have concluded that their organization is the leader in one particular process, they have to justify that with clear evidence. The recommended action ought to be to devise strategies to protect the competitive advantage which has just been identified.

In most organizations, customer involvement is in the design of the product, or the service requirements, but not necessarily in the process itself. Benchmarking presents an opportunity to involve customers more closely in not only looking at the design of the product, but also the design of the delivery process. Perhaps by suggesting how to minimize the number of steps and pointing out to the supplier where the additional operations and features, which are not valued by the customer, can be eliminated from the process. This ambitious way of inviting customer input throughout the business delivery process may be the normal way of doing business in the future. Transparency of partnerships and relationships with the customers will be tested and measured by the extent of using customer involvement.

Transfer of knowledge through secondment may also prove useful. This is the Japanese practice referred to as *shukko*, where employees are encouraged to spend time in competitors and other organizations with a view to sharing knowledge and information on their organizations. They may also learn new practices, new methods and new ways of managing processes which they will bring back to their own organization.

14

How to avoid pitfalls

There are currently more than a dozen benchmarking processes in use by various companies. They all have a similar framework and contain the same basic elements of:

- a **planning or preparation phase** which includes the identification of the subject of the benchmarking study, the identification of benchmarking partners and the establishment of a means of data collection;
- an **analytical phase** which includes the establishment of the benchmark, an understanding of where we are in relation to it (the competitive or performance gap) and an understanding of what is likely to happen to that gap if we do nothing other than what we are already doing, or changes which we have already planned to bring about;
- an **integration phase** which looks at what changes need to happen in order to at least close the gap identified in the previous phase or preferably to overtake and actually become the benchmark; and
- an **action phase** which puts together, implements and subsequently monitors the rolling out of a detailed action plan aimed at bringing about the changes identified in the previous phase.

This chapter looks at the process of working through a benchmarking study and highlights some common pitfalls learned over a long period of time carrying out benchmarking studies on a range of subjects.

A typical study will begin with the identification of the subject to be benchmarked. Although this might sound easy, if a great deal of thought and care is not given to this, then the study is doomed to failure from the start.

The subject of the study may be defined as: a product or service you produce or purchase; a work process or business practice you use; or a factor which is critical to the success of your operation. The important word here is *critical*.

The first part of the preparation phase runs along the following lines. First of all a list of possible subject areas for benchmarking are identified. Some questions which might help with the development of the list are.

- What products or services are you providing to your customer?
- What are the factors responsible for achievement of your customer's satisfaction?

- What decisions need to be made and what information would you like to have prior to making a decision?
- What problems have been identified in your operation?
- In what areas are you feeling competitive pressure?
- What process measurements are you tracking?
- In what areas are your major cost of quality expenditures?

From this initial exercise various list reduction techniques can be used to help select the subject(s) to be benchmarked.

- How critical is it to achieving customer satisfaction?
- How applicable is it to decisions I need to make and how important is the decision?
- How significant is the problem to be solved?
- How important is it to the development of plans and strategy for my area?

Measurements which demonstrate true indicators of performance need to be identified and a written summary of the purpose of the study needs to be prepared which can be shared with the sponsor of the project. This is an important step as it acts as a meeting of minds and ensures a common understanding before any detailed work is undertaken.

Some pitfalls can occur at this early stage.

- **Subject selected not critical** If there is little or no understanding of the link between the subject and results and/or critical success factors, support for the exercise is very likely to diminish to such an extent that it will not reach a conclusion.
- **Too many subjects** If the study has more than one or two subjects at the very most, the effort is likely to be diluted to such an extent that it will not reach a conclusion.
- **Quality, cost and delivery not considered** If only one or two of these elements are considered then the study could possibly affect the one(s) not considered in an adverse manner.
- **Customer satisfaction not considered** Most businesses nowadays are beginning to understand the absolute need of gaining and maintaining customer satisfaction so any exercise not focused in this direction is inadvisable.
- **Too many metrics** Metrics should ideally be the minimum number necessary to capture the key, true indicators of performance.
- **Poor metrics** If metrics are not explicitly tied to true indicators of performance they will make any data analysis difficult, if not impossible.
- **No purpose defined** A brief statement of the purpose of the study is vital to ensure a common understanding between the sponsor and the benchmarking team before detailed work begins and to act as a focal point for the remainder of the project.

- **No management buy in** This is guaranteed to help the study fail at some stage, and should be checked out explicitly to minimize the risk.

Once this has been accomplished, the search for benchmarking partners can begin.

A benchmarking partner could be a company with which you are directly competing, or any company or organization which is considered to be an industry leader in a specific area. This could, therefore, be another division of your own organisation.

If it is not clear from the initial list who are the most appropriate benchmarking partners, it is worth making the extra effort to collect more data on all the companies on your list. Once this is done, ask the following questions.

- How reliable is the information I have about each organization?
- Are there really enough similarities between us and them for them to be partners in this exercise?
- Are there any differences between my business and theirs which may invalidate the results of the study?
- Are they friendly or unfriendly?

This will help reduce the initial list to a manageable number. At this point, there are more common pitfalls to avoid.

- Not considering companies from other industries demonstrates a lack of understanding of the true nature of benchmarking. If your list contains no one from another industry, be suspicious.
- Other departments or divisions within the company are frequently not considered.
- Customers are not consulted.
- Searches of public sources, if often superficial, mainly because people do not know where to look.
- Companies have been removed from lists without adequate information, on the whim of members of newly appointed benchmarking study teams.
- And again, no management buy-in.

The third and final element of the preparation phase is to determine what data is required for the study and how it is going to be collected. First of all a list of questions needs to be prepared. A good starting point for this exercise is to review the purpose, subject(s) and measurements for the benchmark study that were developed earlier. If some questions are difficult to answer for your own operation it may be necessary to modify the question or modify the measurements selected. Answering the questions for your own organization provides a validation of the question and a basis for initial analysis in the next phase of the benchmarking process.

Some common pitfalls during this step are given below.

- **Questions not related directly to the subject** It is very easy to collect information. Unfortunately, unless our efforts are focused very tightly on the purpose of the study, data which is likely to pollute the analysis will also be collected.
- **'So what' questions** Getting answers which do not help us in the analysis probably points to a casual approach in putting the questionnaire together.
- **Questions difficult to answer for your own organization** As previously stated, unless the questions have been answered satisfactorily for your own organization, comparison at any level will be impossible.
- **Inter-relationships and practices not probed sufficiently** The whole purpose of the data gathering exercise is to make accurate comparisons of not just 'Apples and Apples' but 'Cox's Orange Pippins and Cox's Orange Pippins'. Unless this is accomplished then once again any subsequent analysis will prove less than useful.
- **Public sources not used** There is an incredible amount of information in the public domain which can add real value to this part of the exercise for minimum effort.
- **Only one person making company visits** One person cannot possibly act as questioner, note-taker, observer, facilitator and so on. A minimum of two people, with clearly defined roles and responsibilities before any visit is a crucial element of this part of the exercise.
- **Key findings trusted to memory** It is vital that comprehensive notes are taken during a meeting. Notes should be swapped by team members immediately after the meeting. It is also a good idea to tape this note-sharing for future reference.

Once the data has been collected, the next step, the first part of the analysis phase, is to establish the competitive or performance gap. This gap is defined as the difference between your performance and that of the best in the industry and is generally expressed in terms of percentage. The data needs, first of all, to be reviewed to ensure that it is complete and that it is consistent with the questions that were asked during the data gathering process.

Reasons for the gap need to be identified and possible drivers of the gap understood at this point, if subsequent recommendations for change are to be meaningful.

An example of a reason might be: 'The cost gap of 15% is due to salaries being 10–15% higher in our organization and a higher ratio of supervisors to workers (1:20 in our organization versus 1:30 in theirs).'

Drivers of performance gaps are things like business practices, standards and culture. A good understanding of these drivers is important when the stage of establishing new goals and supporting action plans is reached.

Some common pitfalls during this step are given below.

- **Analysis paralysis** It is very easy to analyse every ratio in sight. This can be minimized by making sure that the data collected is only that required for the study and this, of course, must have an explicit link to the purpose of the study.
- **Over precision** It is rarely, if ever, necessary to analyse data to several decimal places.
- **Reasons for gaps not identified** The whole purpose of this step is to understand what the gap is and why it exists. If this is not established then any decisions about changes in current goals would hardly be based on fact.
- **Inter-relationships and/or cause and effects not understood** Enough time and effort needs to be spent to ensure that links between the various bits of data are understood.
- **Not sticking to subject** It is very easy, especially if there is a lot of data, to stray from the subject. Remember that the focal point of the analysis should be the purpose of the study. It is often a good idea to have this narrative visible during this step.

Once the benchmark and the current competitive gap have been established, the next exercise is to try to project what is likely to happen to that gap over time.

The **projected competitive gap** is defined as the difference between your expected future performance and that of the best in the industry. It relates to observed trends and is generally expressed in terms of percentage.

By referring back to the results of the previous step, if data for more than one year is available it may be possible to determine any trends. If data is not available for more than one year then projection can be difficult. However, it is most important to try to understand what is likely to happen.

Try to estimate what will be the benchmark performance level for the next time period. This could be 6–12 months or 3–5 years depending on the topic under investigation. Project your performance over the same period as the benchmark. It is important at this stage to document any assumptions made in preparing the projection.

Since few data points will be available, it is unlikely that sophisticated techniques will be necessary to make the projection. Determine if the gap is widening or closing, and try to estimate the future size of the gap.

There are two common pitfalls encountered here.

- **Over precision** It is not necessary to be over precise. We are simply trying to understand what is likely to happen.
- **Step not completed** If there is a lack of data, it is easy not to bother to complete this step. However, with little or no understanding of what is likely to happen, any recommendations for change are difficult to make.

Once this exercise is complete, the results need to be communicated back into the organization in such a way as to gain acceptance of the analysis by directly impacted management. Once this has been accomplished, work can begin to establish what changes need to be made to current goals in order to bring about gap closure. Once again, approval will have to be sought before change can actually happen.

A common pitfall that can occur here is that the results of analysis are poorly communicated and management and/or customer acceptance are not obtained. It is time well spent making sure that the probability of acceptance is as high as possible. The following outline has proven to be successful.

1. Present the key results and conclusions of the analysis before asking your audience to review detailed data.
2. Be sure that your analysis is based on data and not opinion.
3. Maintain an objective point of view throughout your communication. Becoming defensive or subjective will seriously damage the chances of acceptance.
4. Provide the details of the study as an attachment. The fact that you have detailed data could be one of the most important items in obtaining acceptance for your findings.
5. Your communication (whether it is in the form of a written report or a presentation) should cover the following areas, in this order:
   - an overview of the study which includes key results, conclusions and recommendations;
   - how you went about the study, how you selected benchmarking partners, the method(s) you employed to gather data, the analysis techniques used; and
   - appendices containing the questionnaires used, raw data, other factual evidence.

Once over that barrier, the putting together of a detailed action plan can begin, for which approval would also be needed. As soon as approval is obtained the details of the plan would need to be communicated to all concerned before putting it into place. After implementation, the plan needs to be monitored at two levels. Is it rolling out as per the action plan? If not, why not? Is it actually beginning to close the gap identified during the analysis?

And finally, it is necessary to re-evaluate the benchmarks on a periodic basis to determine if they are still valid. In today's fast moving world any benchmark is likely to be surpassed in a fairly short time. Even at this late stage of the benchmarking exercise, things can still go wrong. Here are some of the common pitfalls:

- action plan not developed or approved;
- action plan not communicated; and

- post implementation not monitored.

If post implementation monitoring and evaluation does not happen then the rest of the exercise will have been a waste of time as it will be impossible to determine the impact of the action plan on the competitive gap identified during the analysis phase of the study.

In summary, then, if discipline is applied to the pitfalls discussed in this chapter the chances of a successful benchmarking study will be increased enormously.

15

Benchmarking in the future

15.1 Benchmarking future competitiveness: is time the most critical factor?

Successful companies are using a potent mix of low cost variety and fast response time to beat competition. The Japanese have established their supremacy by reducing time in all aspects (Stalk and Hout, 1990), including sales and distribution, manufacturing, design and development. By developing strategies which focus on time reduction, great benefits can be achieved.

According to Stalk and Hout (1990), time-based competitors outfox their rivals by adopting the following strategies.

1. **They consider time as the most critical element for establishing competitiveness.** By increasing speed of all activities, reducing delays and eliminating bottlenecks, these companies strongly believe that they will reach their customers and potential customers much more quickly than their rivals. The time advantage can then be used for fine tuning and building ring fences, so that it becomes harder for the competition to persuade customers to switch to their products and services.
2. **They rely on the ability to respond to stay close to customers** by building a big advantage in launching products and services faster than competitors and having a differentiating advantage such as technological uniqueness or other criteria such as image: a concept which the customer values very highly.
3. **They direct the value delivery systems towards the most attractive customers.** This gives time-based competitors the advantage over their rivals in dictating prices, because the most attractive customers are those who are willing to pay higher prices for the right quality standards.
4. **They set the pace for innovation and creativity in their industry sectors.** By launching new products and services earlier than everybody else they have the opportunity to dictate the pace of development. They educate customers to switch to the new products and services and raise their expectations.

   Benchmarking can help organizations develop time-based competition capability by enabling them:

- to examine core aspects of their business (i.e. the right things to do) from the point of view of the customer – they can then concentrate on adding value and delivering quality to the customer much faster than their competitors;
- to develop an obsession with the process, improvement, measurement, the introduction of best practice, so that flexibility, responsiveness and speed can also be optimized;
- to get nearer customers and focus on those who are most sensitive to price and quality; and
- to protect time-based competitive advantage.

Time-based competition is determined by performance measures. These focus on the needs of the customer and establish benchmarks in all aspects of the business delivery process. The following are examples of time-based performance measures used in four critical processes (Stalk and Hout, 1990).

1. **New product development:**
   - time from idea to market
   - rate of new product introduction
   - percent first competitor to market
2. **Decision making:**
   - decision cycle time
   - time lost waiting for decisions
3. **Processing and production:**
   - value added as percentage of elapsed time
   - uptime X yield
   - inventory turnover
   - cycle time (per major phase of main sequence)
4. **Customer service:**
   - response time
   - quoted lead time
   - percent deliveries on time
   - time from customer's recognition of need to delivery

Benchmarking in the future will have to focus not just on quality issues but on time as well. Organizations will have to learn to manage quality in a pro-active manner by optimizing standards in the marketplace through early entry and then fine tuning, refining and optimizing the performance.

As Stalk and Hout (1990) argue:

> Time is a fundamental business performance variable. Listen to the ways in which managers talk about what is important to the success of their companies: response time, lead time, up time, on time. Time may sometimes be a more important performance parameter than

money. In fact, as a strategic weapon, time is the equivalent of money, productivity, quality, and even innovation.

15.2 Establishing a market-driven approach through benchmarking

Unlike the 1970s and 1980s, in the 1990s it has become very difficult for competing organizations to adjust to market pressures and adopt reactive strategies while the business environment is changing. The key factors which led to the radical changes in the marketplace include the following.

- Market segments are breaking and rebuilding at a fast pace, as customers become more and more sophisticated and demand better quality products and services.
- Building competitive advantage nowadays must be looked at in relative rather than absolute terms. Product life cycles are being reduced by up to 50%, entry to markets has become easy through affordable technological capability and the relative ease of copying.
- Threats are coming from beyond national markets, as most competitive markets are becoming global.
- Many markets are being saturated with products and services of similar quality standards and little price differentials. This makes the criteria for customer choice difficult and uniqueness can only come from innovation and a commitment to delight the end customer.
- Customer–supplier relationships are changing, customers are reducing their supply base and developing closer partnerships with the best performers in the marketplace through the exploitation of information technology and electronic communication.
- New management concepts and principles are developing very fast. The pace of change is increasing. Structural and infrastructural changes are more frequent nowadays.

Working harder and harder for the benefit of the customer is no longer applicable. Customers have to be approached in an entirely different way; they have to be involved more closely in the design and development of products and services which they need; they have to be listened to, and communicated with on a regular basis and encouraged to participate in the setting of future strategies. As Lloyd Reuss (General Motors) puts it:

> We have treated the car market as a mass one, but now I am convinced that concept is dead. We now believe in target marketing: specific products and ads aimed at selected groups.

Market-driven organizations are those which put the customer first, align their processes and activities for the optimization of value-added activity, measure performance in customer satisfaction terms and benchmark their practices and degree of competitiveness in customer terms. Peter Drucker (1988) believes that performance results take place outside the organization through ensuring customer satisfaction and not internally as traditionally the case used to be. He argues that:

> Finally the single most important thing to remember about any enterprise is that there are no results inside its walls. The result of a business is a satisfied customer, . . . inside an enterprise there are only cost centres. Results exist only on the outside.

Drucker (1954) also argued that:

> There is only one valid definition of a business purpose: to create a satisfied customer. . . It is the customer who determines what the business is.

15.2.1 Market-driven organization: what is involved?

Market-driven organizations are those that continuously put the customer as the top priority and where the focus is on fulfilling customer needs rather than pushing products and services out in the marketplace.

A market-driven organization has been described as one with a commitment to a set of processes, beliefs, and values that permeate all aspects and activities and are guided by a deep and shared understanding of customer's needs and behaviour and competitors' capabilities and intentions, for the purpose of achieving superior performance by satisfying customers better than competitors (Day, 1990).

Day (1990) presents a comparison of an internally focused business organization against a market-driven one. Whilst the internally focused organization is more concerned with product push, volumes, costs and mass delivery, the market-driven organization starts with understanding customers and their real needs and segments the market according to type of customer. Another noticeable difference between the two types is that in the case of market-driven organizations, knowledge of competitor activity is based on quality information and intelligence rather than gut feelings and subjective opinions as the case is with internally focused organizations.

15.3 Global benchmarking for global supremacy

The establishment of global markets has been greatly facilitated by the exploitation of technological innovation and communication networks. Global competitiveness is based on strategies which integrate national markets by providing the following (Loewe and Yip, 1986).

- **A standardized core product or service.** For example McDonalds and Coca Cola sell standardized products despite individual countries' different tastes. Other examples include the pharmaceutical industry, and semiconductors.

- **Global involvement** and participation in the promotion of products and services to build a global market base, exploit technological opportunities and know how through optimization in all countries concerned.
- A concentration of **value-adding activities** such as R&D, manufacturing to increase economies of scale and exploit expertise, areas of strength for the benefit of the global market.
- **Competitive strategies** designed to create supremacy and total dominance in global markets for specific products and services.

Global competition is very different from the approach used by multinationals. As Levitt (1983) argues:

> The multinational and global corporation are not the same thing. The multinational corporation operates in a number of countries, and adjusts its products and practices in each – at high relative costs. The global corporation operates with resolute constancy – at low relative cost – as if the entire world (or major regions of it) were a single entity; it sells things in the same way everywhere.

Figure 15.1 A model of a global organization.

Figure 15.1 illustrates a model of a global organization (Morgan and Morgan, 1991), where the total design of the organization is geared towards addressing issues such as:

- processes designed and aligned for delivering the needs of global customers;
- resources deployed to deal with global opportunities and focus on core activity;

- structures and communication systems designed to support global activity;
- managing cultural differences through emphasizing the need to focus on the end customer;
- transfer of skills, knowledge and expertise on a global basis, to develop employees and managers with abilities to contribute in global operations; and
- standardization of performance measurement systems, methods and technologies to optimize value added activity on a global basis.

Global benchmarking therefore ensures that superiority is achieved in all areas and amongst all sister companies. Global benchmarking is not just concerned with the specific implementation of best practices in individual operations, it is also about the transfer of learning, and the establishment of best practice on a global basis. Global benchmarking ensures that projects are undertaken according to process strength, availability of opportunities for learning/developing partnerships in 'regional markets' for the benefit of the global organization.

15.4 Competing for superiority – how to beat the Japanese

The challenge for any organization wanting to close competitive gaps against the Japanese lies in its ability to achieve and maintain total customer satisfaction and to have long-term commitment towards its customer base. Benchmarking is the ideal tool to do this. The major distinguishing factor between Western and Japanese companies is commitment to the end customer. In the West, slogans include 'the customer comes first', 'customer is always right'. In Japan the messages are more powerful. They use *okyakasuma wa kamisama desu* meaning 'the customer is God' (Morgan and Morgan, 1991).

Other key factors thought to give Japanese companies a competitive advantage over their counterparts in the West include the following.

- **The corporate family** (*Kaisha*) – where employees and managers work in a close relationship. Matsushita for example uses the following slogan (Morgan and Morgan, 1991):

  Matsushita is a place where we build people.

- **Consensus management**, through a process called *Kessai* (group discussion at all levels, agreements and group decisions).
- An obsession with **quality** and seeking perfection.
- Careful choice of **suppliers as partners** (criteria used includes excellence in facilities and capability, commitment to uncompromising quality, sound long-term strategies, desire to achieve superior standards).

Morgan and Morgan (1991) have produced the following list of factors which is thought to make the Japanese such formidable competitors.

1. Reducing the cycle times for all company activities and functions.
2. The pursuit of continuous improvement in all areas.
3. Total employee empowerment.
4. Bridging customer linkages at all levels, strategic and individual.
5. Ensuring effective use of resources available.
6. Basing product design and delivery on cost effectiveness.
7. The creation of effective communication systems worldwide.
8. Adopting long-term thinking and commitment with short-term plan for incremental results.

15.5 Benchmarking and strategic alliances

As competition intensifies, manufacturers and service providers alike find themselves increasingly having to deploy more resources and expend more effort merely to maintain their position in the marketplace, let alone increase it. Resource availability has become a key issue. In addition, tighter regulations and more attention given to environmental issues has meant competition has become cut throat. Technological advancement and information availability have narrowed the gap between organizations. Managers have come to realize that having a competitive advantage does not mean domination. The competitive advantage has to be protected as much as possible by deploying resources in the right way. So people have started to looked at a new approach to protecting competitive advantage. The concept of partnering or strategic alliances is thus gaining momentum.

Strategic alliances take many forms, e.g. joint ventures, research consortia, cross licensing. Strategic alliances mean there are long-term implications for the partnership, large resource implications and the need for a commitment to manage the interface between the organizations concerned.

Building effective partnerships for benchmarking exchanges creates the opportunity for going beyond specific process exchanges. It means sharing the benefits from joint strategies in a win–win fashion.

There has been much talk about the difficulty in introducing strategic alliances and numerous examples of disastrous outcomes are frequently reported in the literature. In addition, it is sometimes thought that strategic alliances introduce inertia, stifle innovativeness and are not of great benefit to the end customer. Robert (1992), for instance, argues that:

> Alliances generally reduce competition, and reduced competition is not good for the consumer of the companies that are affected. They usually produce higher prices for the consumer and breed complacency in the companies involved.

However, strategic alliances which are developed from a TQM drive should not affect customers in a negative way. In addition, strategic alliances which are born of a culture of exchange through benchmarking create a better climate for innovativeness and creativity because they tackle problems which impede organizations from delivering better quality to the end customer. Lastly, the competitive advantage is going to be unique for each individual organization and may not form part of the strategic alliance agreement. Therefore it will not affect the nature of competition in any significant way.

15.5.1 What is meant by strategic alliance?

Strategic alliances have been described in a variety of ways because they tend to include a wide variety of agreements including joint ventures, R&D collaborations, equity investments, university research agreements, product licensing agreements, etc.

Forrest (1992) suggests the following definition of strategic alliances:

> Strategic alliances are those collaborations between firms and other organizations, both short- and long-term, which can evolve either partial or contractual ownership, and are developed for strategic reasons.

The four key points which can be derived from the above definition is that:

- collaborations may take place not only between firms but also between firms and other organizations such as universities, research institutes;
- alliances can be short- or long-term;
- strategic alliances may involve equity participation or contractual ownership; and
- strategic alliances can input on competitive strategies of the partners concerned and could form an essential element of those.

15.5.2 Learning from Japan: how are partnerships developed?

To understand how strategic alliances and partnerships develop in Japan, it is important, first of all, to focus on the role of the Japanese Government. Essentially, the government in Japan assumes a facilitating role in a pro-active way to promote high productivity levels and long-term competitiveness of Japanese organizations.

The role of the state in Japan can perhaps be examined at three working levels.

1. **Deliberation boards,** referred to as *shingikai* (Ballon, 1992), establish industrial policy that is workable. The composition of these boards includes government ministers and representatives from the industrial

sectors concerned and their remit is to review existing policies and develop strategies for offering help, advice and support.

2. **Administrative guidance** referred to as *gyosei shido* is available on a continuous basis. There is constant dialogue, communication and help lines.
3. **Trade associations** are also very active in Japan. They are a powerful voice representing the views and interest of industry sectors and they bring together the government and industry.

Strategic alliances tend to work in Japan because of the favourable climate created by sound, long-term industrial policies and a state system which is committed to the prosperity of Japan. Strategic alliances work through two main approaches referred to as *keiretsu* (industrial groupings) and *zaibatsu* (interdependence through holding company).

Keiretsu

This system evolved after the second World War when capital was difficult to raise and the only alternative was major banks. Indeed, affiliation to major banks in Japan is the normal practice and banks such as Mitsui, Sumitomo and Mitsubishi played a key role in the development of strategic alliances, over the years. Through the *Keiretsu* system, banks accepted and sometimes encouraged the sharing between companies since there are common benefits.

Japanese centred business alliance known as *kinyu keirutsu* (financial lineage) is also known as *kigyo shudan* (enterprise group) (Gerlach, 1987). The principle is based on co-operation rather than control. There are three critical elements at the core of strategic alliances, as described by Gerlach (1987):

- the creation of high level executive councils representing the group members and also creating a forum for exchange and co-operation;
- the creation of a structure for exchanges and networking to identify the contribution of each specific company and also to identify the constraints and operating boundaries of the alliance; and
- common interest through specific choice of projects that would lead to common benefits, thus demonstrating that the principle of *kigyo shudan* works.

Zaibatsu

The alliance is controlled by a family-held holding company. This principle essentially works through the replication of relationships between suppliers and customer at various levels. Amongst the key factors which make this type of alliance work are the following (Zairi, 1991):

- co-operation and involvement at various stages of the product development process – using joint multidisciplinary teams, problems are tackled and benefits shared; and

- the encouragement of innovative thinking through team work, inter-organizational new product development and management and the sharing of resources, knowledge and other benefits.

In times of hardship and duress, there is often unity reflecting true interdependence, and a common stance. For example, employees who belong to a depressed industry sector are often transferred to alliance partners where the prospects are better and where there is growth. Strategic alliances provide a good equilibrium between external relationships and internal relationships. This therefore enables organizations to focus more on optimizing product development, raising productivity levels and developing skills and knowledge of employees, since there are no distractions from outside and the climate is, generally speaking, supportive and favourable. This happy equilibrium has helped the introduction of lifetime employment (referred to as *shushin koyo seido*) which Japanese society benefitted from in recent times.

This is essentially the way business is conducted in Japan: an emphasis on co-operation and learning through open exchange rather than conflict and obstructiveness. The prosperity of individual organizations is very much dependent on the prosperity of the country as a whole. In relation to this point, Ballon (1992) writes:

> The general acceptance of interdependence between parent companies and their subsidiaries, subcontracting between manufacturers and suppliers, and interdependence of manufacturers, wholesalers and retailers, if not also customers, dictates most of the flow of business. Business organizations and trade associations actively integrate the often divergent interests of individual companies.

Yugoka

Whilst the *keiretsu* are mainly driven by large companies, *yugoka* is a type of strategic alliance which brings together small companies from different industry sectors. The creation of *yugoka* was thought to have started about ten years ago because of pressures on small to medium size companies to meet the ever changing and evolving customer demands. It is thought that there are currently well over 3000 *yugokas* meeting on a regular basis with facilitation and support from the government. The average size of each *yugoka* is 30 companies.

Yugokas were formed as a deliberate government policy to help small to medium size enterprises (SMEs) collaborate, exchange information on areas which are very specific to their types, conduct joint R&D projects and co-operate in all aspects. In 1988, the Japanese government passed a bill to this effect, called 'Bill for temporary measures to facilitate the development of new fields through tie-ups among small businesses in different fields'.

It is left to the Japan Small Business Corporation (JSBC) of the Ministry of International Trade and Industry to manage the working of the various *yugokas*. The criteria for joining a *yugoka* are based on:

- a mixture of industries;
- having similar sizes;
- having similar capital investment;
- only small companies having the capability and know-how to design and manufacture their own products; and
- senior management commitment and the willingness of chief executives to attend and actively participate in all the meetings, which take place once a month.

Overall, government support is given at three distinctive stages.

1. **Knowing each other:**
   - information exchange on common interests, company visits;
   - meetings of small businesses to discuss ideas for joint development;
2. **Using each other:**
   - specific groups with mutual interests are formed and feasibility studies on agreed areas for cooperation are conducted;
   - roles definition for each specific partner;
3. **Creating each other:**
   - the commercialization of joint developments takes place at this stage;
   - partners discuss various possibilities of repeating the experience on another project, or joining other groups or leaving the *yugoka* altogether;
   - it takes 4–5 years for a joint project to be completed.

Progression from stages 1 and 2 is the most difficult aspect and most drop outs take place during those two stages.

SMEs join *yugokas* for various reasons. A survey conducted in 1989 indicated that there are four major reasons why companies wished to join:

1. to exchange information on market and technology (65% of respondents);
2. to improve business performance (60% of respondents);
3. to participate in joint developments of new products (55% of respondents); and
4. to participate in joint development of a new service (35% of respondents).

SMEs do not embark on a full-scale new product development scheme from day one. The survey conducted in 1989 seemed to indicate that most companies in *yugokas* are still at the information exchange and 'getting to know you' stage (60%), 18% of *yugokas* have gone further and are at feasibility stage, and only 14% are at the commercialization stage and maturing.

Furthermore, the survey on *yugokas* revealed that it is very hard to strike perfect partnerships from the beginning. Of all *yugokas* at stage 1, only 27% reported that they were progressing smoothly and 15% are no longer active.

What are the critical factors which lead to the success of yugokas?
What drives *yugokas* is the adoption of best practice technology to derive key benefits and give them the ability to develop new products jointly.

The regular and continuous level of interactions, the concept of sharing information and knowledge on products, services, operations and practices, means that visits have to be organized so companies are exposed to technologies and practices. This is important for success. In addition, representation of participant companies should be at chief executive level, demonstrating high commitment and enabling the partners to move quickly with any decisions that need to be made.

Ultimately, progress that can be made by the various *yugokas* is heavily dependent on the role of an appointed co-ordinator/facilitator, an experienced consultant who is very knowledgeable on technological issues.

The consultant needs to prompt the participants and help them focus on key areas to gather information of strategic value on behalf of all the partners. The co-ordinator/facilitator should highlight potential pitfalls, and point out recent changes in the marketplace to ensure that all participants are up to date.

Good facilitation also means there is continuous learning and conflicts between member organizations are resolved straight away, so that the level of synergy is not threatened.

15.5.3 Examples of successful strategic alliances

Successful alliances often come from partners who have their own competitive advantage or unique strength and who are not directly competing with or after the unique advantage of their partner. Robert (1992) reports on the following two strategic alliances which were undertaken for the right reasons.

3M and Squibb Corp
This is a mutually supportive relationship and each partner brings a key strength. 3M contributes with its strength of polymer chemistry technology that can be used in the context of the development of drugs and which Squibb cannot duplicate. On the other hand Squibb has as a unique strength its distribution system to doctors and drug stores that 3M has no intention of copying.

Apple and IBM
Apple brings its unique graphic and user friendly software whilst IBM uses its powerful RS6000 computer chip. This alliance, it is thought, will bring big leaps in computer technology and will establish the outcomes from this partnerships above other competitors such as Sun Microsystems and Compaq Computer Corp.

Fiat and Peugeot
A strategic alliance took place between the two companies in 1978, referred to as SEVEL (Societa Europa Veicoli Leggeri) for the production of a light van (Lorange and Roos, 1991).

Because of the limited resources, the two companies combined their R&D capability and manufacturing operations. Marketing and sales strategies remained internal to each company however. Because there was no attempt at controlling the course of the alliance, both companies benefited in a variety of ways.

Kubota and SmithKline
Kubota was Japan's number one producer of agricultural equipment for nearly 100 years. However, in 1988, Kubota surprised everyone by shipping its first state of the art and very advanced mini-supercomputer. This success came about through strategic alliances with software and chip manufacturers in the Silicon Valley in the US. Throughout the whole experience Kubota did not attempt to gain control over its partners and kept encouraging them with the provision of resources for the completion of the project.

Kubota used its unique strength of assembly know-how to put the computers together and ship them all over the world (Lewis, 1992). All parties benefited from this partnership: Kubota entered a new market, and the US partners developed further manufacturing know-how through injection of vital capital from Japan, to advance further and protect their competitive advantage.

15.5.4 Examples of failed strategic alliances

Robert (1992) argues that there are mainly three reasons why strategic alliances fail. Each of the following three examples of failures reflect the three pitfalls.

General Motors and Toyota
This was a strategic alliance based on trying to correct a weakness. That meant the stronger partner was allowed to dominate and control. General Motors and Toyota owned a manufacturing plant jointly for the production of small, high quality cars. General Motors entered into this partnership wanting to learn from Toyota on the production of small quality cars since this was its major weakness. Ten years after the project started, Toyota has

performed better than General Motors and seems to be selling six times more cars with its name plate than General Motors.

Interfirst and RepublicBank Corp
This is a partnership of two banks wanting to correct weakness. The new company called First RepublicBank Corp collapsed because the two partners brought with them too many bad real estate loans.

Sony and Bell Laboratories
This type of alliance is basically giving the edge to a partner through the licensing of proprietory technology. Sony managed to acquire its transistor technology from Bell Laboratories for as little as $25 000. This led to the complete removal of all US manufacturers of radios.

Alco and General Electric
The American Locomotive Co. (Alco) faced with a big challenge to stay ahead in its industry and pressurized to look at advanced technologies joined forces with General Electric. The relationship did not work out and Alco was driven out of business as a result of this. Clayton (1992) argues that if one of the partners is bigger in size and if they have full control over key technologies required for the joint project then it is very difficult for the smaller partner to exert any influence at all. If their unique strength is in areas such as marketing, sales, people and customer knowledge, these can be replicated to a large extent. Clayton (1992) concludes by saying:

> The defence of the smaller firm must lie in controlling vital technology that the bigger partner needs to make a saleable product. And regrettably, this Alco never had.

15.5.5 An implementation strategy for effective alliances

Forrest (1992) suggests the following three stages for the development and effective exploitation of strategic alliances.

1. **Pre-alliance stage: matching**
   - right timing of alliance
   - choosing the right partner
   - a good fit between partners
   - each partner seeking an alliance
   - interdependence of partners
   - assessment of strategic costs
   - matching of alliance with overall long-term strategy of firm;
2. **Pre alliance stage: negotiating the alliance**
   - putting in the time and effort
   - strong bargaining position
   - development of mutual strategic objectives
   - involvement of personnel skilled in negotiating

- top management support
- involvement of those concerned
- identification of who will manage alliance
- identification of alliance champions and deal busters
- understanding of partner's strategic intent
- development of trust between partners;

3. **Elements of the alliance agreement**
   - comprehensive detailed agreement
   - scope and objectives of alliance
   - resources to be allocated by partners
   - definition of duties of parties involved
   - how alliance is to be managed
   - patent, intellectual property, publication policies
   - how to resolve conflict
   - milestone points
   - built-in flexibility
   - exit terms.

Following these three stages the alliance should be implemented as follows:

- open communication between the two parties;
- mechanisms in place to facilitate communication;
- focal point person;
- alliance champion;
- mechanisms in place to ensure timely decision making;
- continual mutual commitment of resources needed, e.g. facilities, manpower;
- willingness to change strategic objectives when needed;
- good interpersonal relations between partners; and
- good leadership and motivation of those running the alliance.

Benchmarking is the key for continuous learning and for the development of the urge to constantly improve, adapt and develop strength. As such, benchmarking requires the development of a strategy which addresses issues of choosing partners, the choice of the technology, process or practice that needs to be studied, the desired outcomes and the strategic expectations both short-term and long-term. Although, in most cases, benchmarking tends to be short-term in its workings, the choice of partners is really for specific areas, and the exchange stops at information sharing, and company visits stages, there is scope for extending the exchanges and to go beyond the sharing of information, towards the exploration of joint projects with mutual benefits.

Strategic alliances are a considerable extension of benchmarking and as such enhance the learning process considerably more. Strategic alliances enable organizations to continuously learn and exploit state of the art technologies and practices, they can also extend employee development and thus provide a pool of knowledge that gives a competitive edge to the organizations concerned. On this point, Lewis (1992) writes:

> Highly innovative, 'high learning' organizations have a strong propensity to seek and adapt new ideas from all sources. Every outside contact is seen as a chance to find useful practices to be applied at home. On the inside, high learning firms are woven together by countless informal networks that build collective understandings with new information from experience, partners, customers, classrooms and extensive literature scanning.

Lewis (1992) continues:

> In high learning firms, part of every manager's job includes encouraging experimentation and helping people find better ways to do things. Failure is not only tolerated; it is expected to lead to new insights and add to the organization's knowledge. Assumptions are regularly revisited to avoid the trap of yesterday's wisdom.

These quotes demonstrate the spirit of never-ending improvement and the continuous search for excellence that successful strategic alliances could establish and as such prove that there is full compatibility between the preaching of total quality management and the role of benchmarking in establishing partnerships and seeking best practice for continuous learning.

Part Three

Who is who in benchmarking?

16

Profile of the organizations who have promoted benchmarking

This section gives a profile of the organizations within the UK who are active in the field of benchmarking. The profile reviews the organization, its methodology and relevant work or publications.

16.1 Department of Trade and Industry (DTI)

The DTI ran a competition to assist trade associations in creating benchmarking clubs to help businesses in their sector improve competitiveness. Benchmarking clubs serve to encourage companies to work together in a spirit of self help and co-operation, comparing aspects of their business processes (purchasing, stock control, time to market, etc.) to identify best practice.

Benchmarking helps companies to understand how they can improve their performance. Most large UK industries already have the required resources to undertake benchmarking and many are discovering its advantages. Other companies, especially small- to medium-sized firms would benefit from outside assistance. Trade associations are ideally placed to provide help by setting up benchmarking clubs and encouraging companies in their sector to participate.

The competition 'the benchmarking challenge' helped to finance the setting up and running of around ten benchmarking clubs.

Applications, in the form of a proposal, were needed to identify:

- the key business process to be studied and benchmarked;
- the structure and composition of the benchmarking club;
- the benchmarking processes to be used, identifying any external bodies for this purpose;
- the programme and frequency of meetings;
- the expected costs involved in creating and managing a benchmarking club and the contribution made by the participating companies; and
- arrangements for the wider dissemination of the benchmarking club's report.

Considerable fundraising was made available for the benchmarking challenge and the DTI met some of the costs for the successful applicants.

DTI funding was spread over a two-year period and there was a maximum grant payable to any one benchmarking club.

The DTI continue to support companies with their Enterprise Initiative, offering businesses a co-ordinated programme of information, advice and assisted consultancies on key areas of best management practice. Their free booklet *Best Practice Benchmarking* contains useful and practical advice, and several brief case studies. They also offer sensible alternative sources of information. For a copy ring, 0800 500 200 (UK only). DTI will also provide free advice and help to trade associations not selected for support which decide to proceed with a benchmarking club.

Contact: DTI, Room 5.77, 151 Buckingham Palace Road, London, SW1W 9SS. Tel: 071-215-5847.

16.2 The European Centre for Total Quality Management

The European Centre for TQM is established at the University of Bradford's Management Centre, which is one of Europe's oldest and largest business schools. It occupies an attractive site of thirteen acres of park land in which the buildings are a pleasant blend of the old and new, providing a wide range of accommodation and activities.

The Management Centre pursues an extensive programme of research that has earned it an international reputation, particularly in the field of quality management. It is a fully integrated business school, providing comprehensive and innovative programmes of management education at all levels. The intermingling of students, experienced managers and staff with diverse educational and industrial backgrounds provides an environment that is conducive to creativity.

The TQM teaching, research and advisory services are supported by modern facilities. Short courses for industry are an extremely important part of the work and the Bradford TQM programme has gained international recognition. Several special courses are mounted as part of co-operative training programmes developed with individual companies and professional bodies.

The activities of the TQM Centre are devoted to solving real problems and providing practical methodologies which are transferable. This enables participating organizations to improve quality throughout all the functional areas. Attention is focused on establishing the true requirements at the interfaces of every transformation process and directing effort at meeting those requirements.

The European Centre for TQM is headed by Professor Oakland, the author of *Total Quality Management* published by Butterworth Heinemann, Oxford, 1989 and *Statistical Process Control* (2nd edition, Butterworth Heinemann, 1990). His post is funded by Exxon Chemical and the Centre is funded entirely from contributions of this kind. Hence, Dr Les Porter is the Tioxide Lecturer in TQM, Roland Caulcutt is the BP

Chemicals Lecturer in SPC and Dr Mohamed Zairi is the Unilever Lecturer in TQM. Several other research posts in quality are also funded by external contributions.

16.2.1 Publications of the European Centre for TQM

1. (May 1991) *Benchmarking in innovation within Unilever PLC.* A report commissioned by Unilever.
2. (October 1991) *Understanding innovation management in the FMCG industry sector – Results of a benchmarking study.*
3. *(May 1992) Best practice innovation – A cross industry comparison analysis.*
4. (October 1992) *Best practice innovation: A manager's handbook.*
5. (November 1992) *A competitor intelligence sourcebook in fast-moving consumer goods (FMCG) industry.*
6. (December 1992) *A database on innovation management.*
7. (1992) *Benchmarking in research and development: Managing technical innovation with best practice.*
8. (1992) *Benchmarking consumer needs in FMCG.*

Contact: European Centre for TQM, University of Bradford, Management Centre, Emm Lane, Bradford, West Yorks. BD9 4JL. Tel: 0274 384313.

16.3 Xerox Quality Solutions (XQS)

Since its leadership through quality programme in 1982, over 100 000 Xerox people have been trained to put quality management theories and principles to work. Quality has transformed its business results. Rank Xerox has been the recipient of many prizes and accreditations including the Malcolm Baldrige award, the Deming Prize in Japan.

Building on the experience in integrating quality successfully into its culture, Xerox Quality Solutions (XQS) offers a continuum of seminars, consultancy services, workshops and training to both internal and external clients. XQS uses a core of experienced Rank Xerox employees who have initiated and implemented the Rank Xerox drive for quality.

XQS has developed a modular approach so that the programmes can be tailored to specific customer needs. The XQS training modules are listed as follows:

- quality induction
- documenting work processes
- reducing cycle-time
- manager as a facilitator
- inspecting for quality
- problem solving process

- quality improvement process
- cost of quality
- benchmarking

16.4 Price Waterhouse

Price Waterhouse has undertaken benchmarking through extensive surveys and reviews in industry trends. It is in a position to do this as a result of its large client portfolio. A significant step that Price Waterhouse made in benchmarking was to hold forum meetings to discuss important customer management issues facing large industries. Two hundred people from the UK's largest companies attended these sessions. Interestingly enough, although the participants regarded benchmarking as the second most important topic, fewer than 30% of their companies performed this process.

Contact: Price Waterhouse, Milton Gate, 1 Moor Lane, London EC2Y 9PB.

16.5 Coopers and Lybrand

Coopers and Lybrand commissioned a survey in collaboration with the Confederation of British Industry on the current attitudes to benchmarking within the accounting industry. They conducted 105 interviews with directors drawn from *The Times Top 1000*; 50% of these interviews were with manufacturing companies.

The objectives of the survey were as follows:

- to measure understanding of the term 'benchmarking';
- to find out the proportion of companies which currently use benchmarking;
- to find out the type of benchmarking used and the way the process is carried out;
- to find out the role of internal staff in benchmarking and the extent to which external assistance is used; and
- to find out why the companies which do not use benchmarking have made that decision

Some of the results of the survey were:

- sixty-seven percent of the respondents actively benchmarked processes;
- there were many definitions of the term benchmarking;
- most functional areas of business were adopting benchmarking as a tool for continuous improvement;
- comparison with direct competitors rather than with the class leader was widespread;
- benchmarking programmes were highly successful;
- benchmarking activities were expected to increase; and

- a large majority of non users had yet to discuss benchmarking at the board level.

Contact: Graham Whitney, Coopers and Lybrand, Benchmarking Centre of Excellence, 1 Embankment Place, London WC2N 6NN. Tel: 071 583 5000.

16.6 Oak Business Developers

Oak Business Developers advocates that benchmarking should be seen in the context of TQM and problem analysis. During the latter half of 1990, it carried out research on the views and experiences of senior managers in UK organizations.

The chief objective of the survey was to establish current levels of awareness and potential growth in the use of best practice benchmarking. A secondary objective was to gather current perceptions of senior managers on best-in-class companies to serve as a starting point for establishing benchmarking networks in the UK.

As best practice benchmarking was little known as a management technique in Britain, the questionnaires were personally handed to senior managers rather than mailed out. The respondents represented an impartial and random sample of practising managers contacted either through a network of business associates or through attendance at courses at a leading Management Development Centre.

It became apparent, from comments provoked by the survey, that to British managers this is a novel and stimulating topic. Response reflected the general lack of awareness surrounding benchmarking. This was further borne out by specific findings of the questionnaire.

Contact: Oak Business Developers, Long Gables, Templewood Lane, Farnham Common, Bucks, SL2 3HJ.

16.7 British Quality Foundation (BQF)

The British Quality Foundation is the focus in the UK for national and international activities in the field of quality, and provides for the interchange of information on quality topics amongst its members. It is committed to furthering the interests of all its members through the advancement of quality competitiveness and hence promotes benchmarking. The BQF supports DTI and other government quality initiatives. It has strong links with the European Foundation for Quality Management and other organizations concerned with the promotion of total quality practices and it is represented on standard developing bodies through the participation of its members.

16.8 The Benchmarking Centre Ltd

The Benchmarking Centre has been formed by some leading international organizations who see a clear need in Europe to promote and facilitate benchmarking. It exists to:

- help to identify and qualify benchmarking partners and facilitate the exchange of information;
- provide common definitions and procedures within an agreed ethical and legal framework;
- promote 'best practice' as recognized internationally; and
- network with existing and future national and international groups involved in benchmarking.

Contact: The Benchmarking Centre Ltd, c/o Dexion Ltd., Maylands Avenue, Hemel Hempstead, Herts HP2 7EW. Tel: 0442 250046.

16.9 European Foundation for Quality Management (EFQM)

In recognition of the potential for competitive advantage through the application of total quality, fourteen of the leading western European businesses took the initiative of forming the European Foundation for Quality Management in 1988. By October 1992, membership had grown to over 230 members from most western European countries and business sectors.

EFQM has an important role to play in enhancing the position of western European businesses in the world market. This can be achieved in two ways:

- accelerating the acceptance of quality as a strategy for global competitive advantage; and
- stimulating and assisting the deployment of quality improvement activities.

The EFQM does not address the process of benchmarking directly. Benchmarking is addressed indirectly through the European Quality Award Scheme. Applicants are given feedback about their performance in various areas in relation to what is considered best practice.

Contact: European Foundation for Quality Management, Building 'Repal', Fellendord 47a, 5612 AA Eindhoven, The Netherlands. Tel: +31 40 461075.

16.10 International Benchmarking Clearing House

International Benchmarking Clearing House was launched by the American productivity and quality centre (APQC). Eighty seven organizations participated in the design of the clearing house. They all recognized the value of benchmarking as a TQM tool and believed that a co-operative effort would create a unique resource for improvement.

The Clearing House aims to provide the following:

- fast and simplified access to domestic and international benchmarking information;
- substantial time and cost savings through reduced staff time and travel expenditures;
- opportunities to network with benchmarking professionals;
- assistance in screening potential benchmarking partners;
- a valuable source for information on business processes; and
- support for meeting benchmarking requirements for the Malcolm Baldrige National Quality Award.

The APQC have instituted several benchmarking prizes under the auspices of the clearing house:

- *Benchmarking Research Prize*
  This aims to extend the body of benchmarking knowledge through the recognition of innovative contributions, developments and research in techniques, methods and tools of benchmarking.
- *Benchmarking Study Prize*
  This aims to promote, encourage and recognize the execution of a benchmarking study.
- *Award for Excellence in Benchmarking*
  The purpose of this award is to recognize those organizations which demonstrate the consistent application of benchmarking according to the award criteria and whose practice of benchmarking represents both leadership and innovation in their use of this process for business process improvement at both strategic and tactical levels.

Contact: International Benchmarking Clearing House, c/o American Productivity and Quality Centre, 123 North Post Oak Lane, Houston, TX 77024-7797, USA. Tel: 713 685 4604.

Suggested reading

Watson, Gregory, H., (1993)*Strategic Benchmarking: How to Rate your Company's Performance against the World's Best*, John Wiley & Sons Inc, New York.

16.11 Profit impact of market strategy (PIMS)

PIMS is an international business consultancy, based on uniquely objective methods of strategic analysis. They work with clients to:

- determine the underlying potential of each business by benchmarking its performance against strategic peers who are similar in terms of

market share, investment intensity, quality position, customer profile, rate of innovation etc;
- validate which tactics have actually worked for business in the same competitive situation.

PIMS reviews approximately 200–300 aspects of a business. This is compared to similar company profiles kept on a database. This database represents 22 000 business years. It is updated regularly and hence becomes more statistically valid. All information is sorted and becomes anonymous once it is entered into the database.

Benchmarking is used for four different tasks – diagnosis, target setting, tracking and forecasting. The PIMS approach reviews:

- *Par benchmarks*
  This is based on real evidence on what variables affect a particular dimension of performance. Quantified effects of factors like customer complexity and innovation are noted. The performance information on normal business trends is noted.
- *Business benchmarks for target setting and forecasting*
  These are used to answer 'what if' situations when forecasting strategies or target setting.
- *Business benchmarks for tracking*
  This reviews whether a company's shift in performance is normal or detrimental in line with market performance.

In addition to business performance, PIMS also looks at a company's quality strategy. It benchmarks a company's product conformance, proper specifications, effect of transport on quality, competitive influence, customer service and customer satisfaction.

Contact: PIMS Europe Ltd., 56 Haymarket, London SW1Y 4RL. Tel: 071 930 5055.

16.12 Personnel Euro-club (PEC)

This Brussels-based organization is a non-profit organization which undertakes various activities in the field of human resources. Its aim is to build a network between European personnel directors to exchange experiences in human resources management in an authentic and thorough manner.

PEC offers its member companies social and economic effectiveness through better integration of human resources management into their development strategy. This is achieved through co-operation between human resource managers and the heads of important centralized departments.

Contact: Personnel Euro-club, 44 Rue Washington, B-1050 Brussels, Belgium. Tel: 32 2 648 4769.

16.13 The Benchmark Partners Inc.

Offers a practical two-day benchmarking workshop built around a year long study of the best benchmarking companies and practices in the country. Also offers the National Quality Survey, a user owned, software driven, Baldrige Award oriented quality assessment process that provides ongoing detailed measures of organizational quality strengths and weaknesses.

16.14 The Inter-company Productivity Group

The Inter-company Productivity Group is a private, voluntary group of leading UK industrial companies. Its purpose is to encourage the sharing of good management practices to improve international competitiveness and to help develop the sorts of strategies and practices needed for the future.

Its activities are decided upon and steered by a strategy group composed of manufacturing and personnel directors from the member companies. The organization first began giving attention to improving productivity via changes to working practices. After conducting a survey, the member companies formed a strategy to pool their experiences of best practices.

Since then, working groups have been formed to explore issues that focus on manufacturing strategy, managing change and working practices. Future activities will look at the following areas:

- role, competence and development of manufacturing managers;
- total quality management;
- leading edge management;
- implications in the developments of information technology;
- supply chain management; and
- working practices.

16.15 IMD Lausanne

IMD Lausanne, Switzerland, operates the manufacturing 2000 programme. Part of this is to benchmark the manufacturing practices across Europe and to make the results available to their members.

16.16 IBM Consultants

IBM Consultants offers a free benchmarking consultancy service to UK industry. Its initial 250 surveys have been reported in the *Made in Britain* report. It is extending work into the engineering sector and there is a growing database becoming available for benchmarking UK companies.

16.17 IFS International

IFS International has established a UK best practice club to compare best practice across UK industry and make it available to its members. Much of this experience is based on its operation of the DTI *Inside UK Enterprises* activity.

17

A review of key publications on benchmarking

17.1 Introduction

This section provides details of six publications.

1. *Benchmarking – the search for industry best practices that lead to superior performance*, written by Robert C. Camp and published in 1989 by The American Society for Quality Control Press, Milwaukee, USA.
2. *Benchmarking – a tool for continuous improvement*, written by C.J. McNair and K.H.J. Leibfried and published in 1992 by Harper Business, New York, USA.
3. *Best practice benchmarking – the management guide to successful implementation*, written by Sylvia Codling and published in 1992 by Industrial Newsletter Ltd., Bedford, UK.
4. *Strategic benchmarking – how to rate your company's performance against the world's best*, written by Gregory H. Watson and published in 1993 by John Wiley, New York, USA.
5. *Benchmarking – a practitioner's guide for becoming and staying America's best of best*, written by Gerald J. Balm and published in 1992 by Quality and Production Management, USA.
6. *Competitive benchmarking – an executive guide*, written by Dr Mohamed Zairi and published in 1992 by Technical Communications (Publishing) Ltd., Letchworth, UK.

This is followed by an abbreviated list of articles on the subject of benchmarking.

17.2 Benchmarking – the search for industry best practices that lead to superior performance

Robert C. Camp begins with both the definition of benchmarking as provided by D. T. Kearns, CEO of Rank Xerox and then his own working definition of benchmarking as 'the search for industry best practices that lead to superior performance'. Throughout the text, the author supports the description of the benchmarking process with references to the experiences of Xerox when it undertook to benchmark its logistic operations against those of L. L. Bean. The author provides an overview of the Rank Xerox ten-step process for performing benchmarking. This framework represents the methodology promoted and provides the basis for the structure of the book.

Camp outlines potential sources of information – both within and external to the organization – which are necessary inputs to the benchmarking process. He discusses the purpose for and reasons why organizations should benchmark, and benefits are identified. The author identifies what to benchmark; the necessity of clearly defining and documenting the organization's own operations and processes before benchmarking against others (either internal or external to the organization) is emphasized. The task of deciding the type of benchmarking to be undertaken, i.e. internal, competitive, functional, or generic benchmarking, and the issue of identifying and selecting potential benchmarking partners are also discussed. Priorities for developing information sources are discussed together with a review of the potential problems which may be encountered in obtaining relevant information from direct competitors.

The criteria for information gathering and the differing sources of information is reviewed, with a discussion of the importance and relative merits of public domain information, trade association data, and using consultants as a source of information.

The importance of building a relationship with benchmarking partners through the sharing of information is highlighted, as is the need to agree a basis for such sharing. Construction of questionnaires is also considered. Camp stresses the importance of determining the qualitative nature of the competitive gap before assessing its quantitative extent, as this approach reveals the activities which support superior performance. The process of projecting future trends is reviewed, and the 'Z' chart is introduced as a tool to display, graphically, historical and current performance, together with the necessary improvement required to match the best practices. The importance of communicating the results of the benchmarking process within the organization at all levels is stressed.

The issue of setting goals based upon the results of benchmarking are discussed and Camp notes that it is likely that organizations may find that future performance objectives require the introduction of new metrics. A key feature of goal setting, he stresses, is that benchmarking is not based upon the extrapolation of past performance but instead relies on an understanding of the organization's external environment.

The next stage of the benchmarking process described is the development of action plans based upon the results of the process. The author emphasizes the need to embed benchmarking within the planning process and support the development of benchmarking as a way of life within the organization. Two facets of the implementation phases are identified: the first involves developing plans related to the tasks and activities to be performed by the employees; whilst the second deals with the culture of the organization and ensuring understanding and acceptance of the plans.

Camp supports the necessity of performing an 'analysis for implementability' (a statement indicating what is to be changed and where responsibilities will be placed) and an 'analysis of activities' (determining the capacity of the process to handle all the required transactions).

Camp emphasizes that functional management is ultimately responsible for the planning and execution of benchmarking practices. Four approaches to implementation are discussed, these being: by line management; by project or program teams; by a Process Czar; and by performance teams.

The author recognizes the need for an inspection procedure, to determine both that results have been implemented and that the benchmarking process was performed correctly, and considers the final stage in the benchmarking procedure, recalibrating the process to ensure that the benchmarks are updated and meet the current industry best practices.

The author finishes with a brief discussion concerning what lies beyond benchmarking and assesses the benefits to the organization from business simplification.

17.3 Benchmarking: a tool for continuous improvement

McNair and Leibfried stress the links between benchmarking and the increasing need for organizations to have an external focus. In a free market economy, stakeholders have other investment opportunities, so for an organization to succeed in the long run it must expand its focus to ensure it is able to respond to the needs of all stakeholders. Also emphasized is that benchmarking is an ongoing process leading to continuous improvement, and the necessity of management having a pro-active perspective to their activities.

The need to undertake an assessment of internal processes before benchmarking against others is discussed. Although the benchmarking process may result only in modifications to existing practices, the authors recognize the occasional need for radical restructuring to achieve 'quantum leaps' in performance.

It is noted that benchmarking can focus on roles, processes or strategic issues in seeking best practice and helps to identify those features on areas which are critical to the success of an organization.

A brief outline is given of the benchmarking process.

The necessity of first identifying and mapping the key processes is emphasized with the application of work flow analysis to link tasks through and across functional boundaries. A key benchmarking task discussed is the development of a 'generic' activity grid, which involves identifying those activities common to other internal functions or external organizations for which a direct comparison can therefore be made.

With regard to data collection, the importance of careful construction of questionnaires is highlighted.

Completion of the analysis phase provides an implementation plan for process change; the authors suggest an 8-step implementation sequence, and each individual stage is then reviewed in greater detail. The relationship of benchmarking to the continuous improvement philosophy and radical process redesign is also discussed in greater depth.

A number of approaches to benchmarking with external organizations are considered, and the relative merits of competitive benchmarking and industry benchmarking are discussed. Likely problems to be encountered in gaining information from both competitors and other companies in the same industry, and other potential sources of information are discussed.

The problems associated with the traditional functional structure of organizations compared with the horizontal value chain which actually meets the customers' requirements is reviewed in depth. The deficiences associated with the traditional focus on financial measures are highlighted and the need to provide qualitative measures which describe the underlying processes are discussed. The authors highlight that true performance is only achieved if an organization is able to move forward in quality, productivity, delivery and cost.

The authors discuss the characteristics of best-in-class benchmarking, with particular emphasis on moving beyond common industry organizations to identify those with superior performance in generic operations. Also discussed are the problems encountered within organizations in assimilating and accepting the results of the benchmarking analysis and the need to create culture which actively supports learning and creativity.

The discussion is then extended to the issue of the identification of the performance drivers of the organization. It is emphasized that such a process begins with an understanding of the internal strategies, structure, working practices and constraints that define the organization's processes. The necessity of correctly identifying these drivers for the appropriate selection of benchmarking partners is noted, as is their use in determining benchmarking measures.

The importance and place of the employee in the improvement of performance is discussed. Clear goal setting and the provision of a means of assessment are identified as critical to the success of any performance improvement programme.

The authors conclude by noting that benchmarking is a process of raising awareness within an organization and developing a culture that is willing to learn.

A number of case studies are provided throughout the text to provide examples of the practical implementation of benchmarking.

17.4 Best practice benchmarking: the management guide to successful implementation

Sylvia Codling begins with a brief historical review of benchmarking, highlighting the various applications of the term and noting the pioneering work in the area of the benchmarking management practice by Xerox. The aims and objectives of organizations such as the British Quality Association (BQA), the European Foundation for Quality Management and the UK Government through the Department of Trade and Industry are noted.

The author states that the traditional competitive analysis has focused at most on those organizations within the same sector and has failed to identify the superior performance achieved by others in alternative industries.

Codling segments the benchmarking process into four stages: planning; analysis; action; and review. These are then split further into twelve steps. Each of these steps is discussed in depth, with guidelines being given for the selection of suitable subject areas for benchmarking and the identification of data sources.

Three distinct types of benchmarking are identified: internal; external; and best practice. Internal benchmarking is defined as benchmarking with partners from within the same company, or division. The relative ease of access to data and the similarities in culture suggest that this approach may prove most appropriate as the first step in benchmarking for an organization. External benchmarking relates to performance comparison with partners from differing business units of the same organization, or with different companies. The author notes that the emphasis on external focus intrinsic to this approach offers greater potential for identifying superior performance and for rectifying cultural opposition to the adoption of ideas from outside the organization. It is recognized that attempts to benchmark with direct competitors is likely to prove difficult due to the issues associated with sharing sensitive information. Best practice benchmarking involves identifying and comparing performance against the owners of processes regarded as best in best-in-class. The author indicates that an initial problem will be to determine a definition for 'best'. It is suggested that only through correct and thorough planning, and data collection can this be achieved.

The increasingly rapid rate of innovation, together with the growing interdependence between organizations and their customers are identified as major drivers for the need of companies to focus to a greater extent on the external environment. The development of organizational structures which support greater responsiveness to customer needs are discussed as is the growth of the TQM philosophy. It is noted, however, that the capacity of problem-solving teams to deliver performance improvement will deteriorate unless reference is made to the external environment. It is in this role that benchmarking acts as a means to provide insights into new practices.

The benefits of an established TQM culture to support the benchmarking process are reviewed, with particular emphasis on the existence of teams with a clear perception of the organizations' objectives and a reduced level of resistance to change.

When choosing an appropriate subject to benchmark, the author suggests that organizations should focus on those areas critical to the business's success. In presenting an overview of the benchmarking process it is indicated that the overriding characteristic of the benchmarking process 'is the discipline it requires and imposes'. Organizations should be aware that the objective of achieving competitive superiority can only be realized over the longer term. Although the nature of the benchmarking process may vary for each organization, the need for vision, commitment and diligence are recognized as essential.

The issues involved in building a constructive relationship with partners are also discussed. The importance of assessing both qualitative and quantitative data is reviewed, as is the need to set targets which result in a narrowing of the performance gap, whilst recognizing the capacity of partners to achieve further improvements in their own performance.

Having presented the planning and analysis phases, the author progresses to the implementation phase, stressing that the essential requirement for successful implementation is the total commitment from all those involved in the process. The importance of communicating the findings of the benchmarking process are also stressed. The author indicates the importance of ensuring that goals and objectives are flexible so that the organization can respond to changes in its environment.

Many companies have achieved superior performance, only to see their competitive advantage eroded. The author discusses the need for companies to re-calibrate their activities to ensure their continued validity with respect to external environment. To provide continuous support for the benchmarking process within the organization it is suggested that a Business Review Executive (BRE) be appointed, who with a thorough understanding of the organization's practices and processes, would be responsible for seeking out and identifying best practice elsewhere, and for promoting benchmarking activities within the company.

Throughout the text, real examples are used to support the arguments presented by the author. In addition a chapter containing more detailed case studies from industry is provided.

17.5 Strategic benchmarking: how to rate your company's performance against the world's best

Gregory H. Watson begins with a brief introduction to benchmarking and recommends a four-step approach.

1. planning study;
2. conducting research;
3. data analysis; and
4. adapt, improve and implement findings.

It is suggested that for an organization to compete successfully it must emphasize: quality beyond that of its competitors; innovation ahead of competitors; and achieving costs below those of competitors.

The author introduces a model for business co-operation which adds the fifth 'P' of 'process' to the traditional four marketing 'P's of product, price, promotion and place. It is suggested that benchmarking represents a fundamental shift in the competitive philosophy, with companies sharing information to reduce development times and increase capabilities.

Watson discusses linking strategic planning with benchmarking and then progresses to the integration of the organization's strategic intent, together with the development of core competencies and process capabilities, as a means to produce a competitive map of an industry.

In his explanation of the essentials of process benchmarking, the author notes that the foundation for successful benchmarking is 'the appropriate acquisition and application of quality methods combined with the continuous pursuit of business knowledge'. Before an organization attempts a benchmarking exercise, it is recommended that it undertake a quality maturity assessment to determine its capability to effectively perform such an activity. The quality maturity exercise can be defined in terms of a four-stage process consisting of inspection, control, partnership and maturity. It is indicated that a limited form of benchmarking process can begin as early as the control stage.

The theory and principles of benchmarking are reviewed with a methodology consisting of four principles provided as the basis for conducting a study: reciprocity, analogy, measurement, and validation of an organization's process measurement. This discussion is followed by a description of the benchmarking code of conduct which includes issues such as principles for the exchange of information and confidentiality.

In establishing a methodology to approach the process of benchmarking, reference is given to the benchmarking template developed jointly by Boeing, Digital Equipment Company, Motorola and Xerox.

The author notes the importance of focusing on and benchmarking the key business processes, as improvements in these areas offer the greatest potential for reward.

Watson recommends the widely used TQM metrics of first-pass yield, value-to-cost ratio and cycle time to assess effectiveness, process economy and efficiency respectively.

With reference to analogous processes, the necessity of applying the same measurement and analysis procedures to benchmarking partners as used within the organization is highlighted.

The author discusses in depth each of the four stages of the benchmarking processes. Three phases are identified in the planning stage: the identification of strategic intent, core competencies, capability maps, key business processes and critical success factors; documentation and analysis of the process to be benchmarked; and criteria for benchmarking partners.

For the second stage of data collection, a further three phases are identified: internal data collection; secondary research; and external primary research.

The use of gap analysis to illustrate the variation in performance between the organization and its benchmarking partners is introduced as is the setting of goals and objectives for achieving performance improvement.

The author debates the issue of learning within companies, addressing the issues of how long organizations take to learn and how to support corporate learning, and gives a brief overview of potential sources of knowledge in relation to benchmarking.

To illustrate the implementation of the four types of benchmarking (internal, competitive, functional and generic) the author provides detailed case studies from Hewlett-Packard, Ford, General Motors and Xerox. The author also demonstrates the applicability of benchmarking to the service industry by following these four cases with three service-oriented cases from International Facilities Management Association, First National Bank of Chicago and the Healthcare Forum.

The concluding chapter of the book provides advice from those responsible for initiating benchmarking within their organizations. Common problems are identified and key strategies to be pursued are highlighted.

17.6 Benchmarking: a practitioner's guide for becoming and staying America's best of best

Gerald J. Balm's book is a practical and effective guide to benchmarking and how to apply this tool within an organization. As an employee of IBM at Rochester, Minnesota, IBM examples are cited throughout the book. Balm frequently cites Robert Camp's *Benchmarking – the search for industry best practices that lead to superior performance*, and relies heavily on information contained therein. Balm's book does not substitute for Camp's book and should be used in conjunction with it. Balm's book continues Camp's work, and lays out a 15-step process used by IBM

Rochester in their benchmarking experiences. These provide a decent guide to this important quality tool, and his concise writing style clearly defines both benchmarking and the methods for its application.

17.7 Competitive benchmarking: an executive guide

Dr Mohamed Zairi's report was the first European guide to benchmarking, and received considerable attention in the management and trade press and national newspapers. The publication is aimed at busy executives who need to get to grips with benchmarking quickly and easily, with the minimum of effort, and while short on theory it is long on practice, with as much case study material as possible included.

The report discusses the background to benchmarking in terms of competitive advantage, and defines the term benchmarking, reviewing the reasons for undertaking benchmarking. The relationship between benchmarking and TQM is also discussed. The different approaches to benchmarking are discussed using examples from Xerox, AT&T and Alcoa.

It includes a review of the different types of benchmarking: internal, competitive, functional and generic. The relative merits of cost-driven and process benchmarking are also analysed.

The relationship between benchmarking and measurement and understanding the competition is considered, as is the role of quality costing, SPC, measurement for effectiveness (the role of performance measurement) and measuring competitiveness (the role of benchmarking). The issues examined include the role of competitive analysis and monitoring competitors' behaviour. The author looks closely at the power of economic data (such as profitability ratios, revenue ratios, productivity ratios and value-added ratios). In particular, the Indicator for Commercial Competitiveness is examined.

The key role of customers is examined, in particular the role of the customer in terms of evaluating quality standards, as a provider of information on competitors and as a continuous source of information. As an example, the role of customers in determining performance standards in the implementation of advanced manufacturing technology is included.

The various tools used in benchmarking are examined. Issues covered include: quality function deployment (QFD), the role of the ISO 9004 framework and the Malcolm Baldrige Award.

Zairi examines the development of an effective benchmarking strategy, including how to understand the business process, using TQM as a competitive weapon and linking benchmarking to the never-ending improvement ethos, and a practical implementation plan is presented which covers the various approaches to benchmarking and the implementation steps. The pitfalls and review systems of benchmarking are also examined, and a detailed case study of benchmarking in innovation is presented.

17.8 Articles

Altany, D. (1990) Strategies: copycats. *Industry Week*, **239** (21), 11–18.

Benchmarking is the formal process of measuring and comparing a company's operations, products, and services against those of top performers both within and outside that company's primary industry. In the corporate world, this type of copying is not only legal and ethical, it is virtually mandatory for a company that hopes to be considered a world-class competitor. Benchmarking's benefits as a strategic planning method are: it identifies the keys to success for each area studied; it provides specific, quantitative targets; it creates an awareness of state-of-the-art approaches; and it helps companies cultivate a culture in which change, adaptation, and continuous improvement are actively sought. It is the subtle way in which benchmarking changes a company's culture and mentality that leads to the greatest long-term gains.

Altany, D. (1991) Share and share alike. *Industry Week*, **240** (14), 12–17.

To achieve the information edge, leading companies employ a technique called benchmarking. In its simplest form, benchmarking is a process companies use to methodically track down business practices and approaches judged to be among the best in the world. The essence of benchmarking is to seek out, learn, and incorporate new operational approaches by exchanging information with top-performing noncompetitors. Benchmarking expands most companies' competitive focus because it delves beyond the end-product analyses that companies typically concentrate on and explores the underlying processes that go into producing those products or services. Expert benchmarkers assert that targeted benchmarking strongly enhances breakthrough thinking by establishing an environment and a network of efficient processes that does not squelch creativity. This focus on continuous improvement and the pursuit of excellence is the feature that gives benchmarking such power when put into practice.

Altany, D. (1992) Benchmarkers unite: clearinghouse provides needed networking opportunities. *Industry Week*, **241** (3), 25.

The concept of a comprehensive benchmarking training, networking, and data storehouse became operational in February 1992 at the American Productivity & Quality Center (APQC). Properly conducted benchmarking studies virtually assure tangible gains, but they are expensive and difficult to perform. For example, companies have trouble identifying top-performing companies in specific functions and finding companies that have already conducted benchmarking studies in specific areas. The benchmarking clearinghouse, with strong support from the US' top benchmarkers and a corporate-membership roster composed of many Total Quality Management companies, will serve as a central networking locale. APQC helps bring people and institutions together to pursue benchmarking studies. In addition, the clearinghouse will provide online access to abstracts of companies' benchmarking studies, as well as data derived from the studies.

Anonymous. (1992) How three companies use benchmarking to meet corporate goals. *Environmental Manager*, **4** (5), 6, 16.

Shortly after it was founded in 1991, Rhone-Poulenc Rorer embarked on an international environmental benchmarking project. The benchmarking process enabled the pharmaceuticals manufacturer to achieve several objectives. One was the ability to prioritize its corporate environmental agenda to focus first on the plants with higher environmental index scores. A joint effort by AT&T and Intel has been aimed externally at benchmarking pollution prevention. A nine-step process was used: identifying the bench; planning; secondary research; deciding which companies to focus on; primary research; data analysis; implementation planning; implementation; and calibration. Brenda Klafter of AT&T Bell Laboratories sees benchmarking as a powerful, team-based tool that can help in the development, improvement, or reengineering of a process or a program. With management support, team buy-in, resources, momentum, and enthusiasm, companies can create positive change using benchmarking as a tool.

Anonymous. (1993) The benchmarking boom. *HR Focus*, **70** (4), 1, 6+.

Human resource management (HR) can play a pivotal role in benchmarking, both in making sure that employees are trained properly to benchmark and in conducting its own benchmarking studies of human resources functions. Federal Express is one benchmarking leader that has been flooded with requests. The company gained its notoriety after it won the Malcolm Baldrige National Quality Award in 1990. Effective benchmarking requires looking at competing companies as well as at companies that are not in that particular industry but have the same or similar processes or functions. In terms of areas that tend to be benchmarked most often in HR, Linda Crosby DeBerry says that Federal Express's leadership evaluation and assessment process receives a lot of attention, as does the company's survey/action/feedback program.

Anonymous. (1992) Benchmarking environmental audits: how does your company measure up? *Environmental Manager*, **3** (12), 3–4, 12.

In an interview, Larry Cahill of McLaren/Hart Environmental Consulting, an expert on environmental audits and author of the new, sixth edition of Environmental Audits, offered a number of observations and suggestions on environmental audit programs. A good choice for an environmental auditor is a person who can walk into a plant site and feel comfortable and who has good interpersonal skills, along with an understanding of regulations. The primary job of an auditor is to obtain information. The most important skills that environmental auditors need are communication-based skills that include: how to interview and listen; how to write a report so that people can understand it; and how to get the message across. Auditors use interviews, inspections and looking at records to gather information. One of the biggest problems with environmental audit programs is following up with corrective actions. The entire audit process can break down because of poor follow-up.

Anonymous. (1993) Using a maintenance benchmarking study. *Pulp & Paper*, **67** (2), 48.

In a panel discussion, Taylor Heidenheim of Weyerhaeuser Paper Co., Jim Grant of Weyerhaeuser Paper, and Christer Idhammar of Idcon Inc. discussed the value of maintenance benchmarking, with the main points focusing on what benchmarking is, how paper mills can benefit from the data, and certain pitfalls of analysing data.

Anonymous. (1992) The benchmarking 'attitude'. *Environmental Manager*, **4** (4), 10.

Originally a surveyor's term for a reference point, the word benchmark implies a standard by which something can be measured or judged. However, Alan Gagnet of Environmental Quality Corporation EQC refers to benchmarking as an attitude oriented toward continuously making broad incremental improvements across the board. Benchmarking is an attitude that is in accordance with the recommendation of the Environmental Protection Agency's Science Advisory Board that what is needed to protect the environment is a new mindset. Any company with more than one site should start the benchmarking process internally. It is important to keep an open mind when doing environmental benchmarking. Part of the benchmarking process is that the organization accepts that a better way to do things exists.

Anonymous. (1993) Benchmarking aids in benefits design. *Employee Benefit Plan Review*, **47** (12), 60–61.

In the benchmarking process, an individual or company learns from another in order to continuously improve. Through benchmarking, an individual or company finds out what techniques a role model uses that could work for it. 3M Co. used benchmarking to change its employee benefits programmes. 3M's benefits benchmarking methods include surveys, face-to-face interviews, and consultants.

Anonymous. (1993) Six steps to benchmarking success. *Traffic Management*, **32** (4), 64.

Alcoa and its chief executive officer view benchmarking as an integral component of the total quality management process. The company has been aggressively seeking out best practices since 1987. Figuring prominently in the benchmarking activities is Alcoa's procurement and logistics function. The department has benchmarked the entire procurement process as well as specific transportation and warehousing operations. Alcoa believes in a formalized approach to benchmarking and has developed a six-step methodology that guides managers through the process. Alcoa's six steps are:

deciding what to benchmark; planning the benchmarking process; understanding the company's own performance; studying others; learning from the data; and using the findings.

Balm, G. J. (1992) Benchmarking – nicety or necessity? *Tapping the Network Journal*, **3** (1), 6–8.

Combined with process management and project management, benchmarking presents a powerful capability to efficiently and effectively make dramatic improvements. Process management helps companies prepare to benchmark their most important and highest payback processes, products, and services. Project management assures that improvement ideas generating from benchmarking will be effectively implemented and monitored. Although benchmarking has existed for many years, it has enjoyed a rebirth as a result of an expanded scope pioneered by Xerox, Motorola, IBM, and others. Benchmarking received considerable recognition for its Six Sigma defects goal. As defined by Motorola, this is a defect prevention-based system that allows no more than 3.4 defects per million opportunities. Using Motorola as the benchmark, IBM adopted a similar defect reduction program. Deployment factors that were pertinent to IBM's effort included top management support and proper preparation before benchmarking.

Band, W. (1990) Benchmark your performance for continuous improvement. *Sales & Marketing Management in Canada*, **31** (5), 36–38.

Every company should regularly evaluate or benchmark its performance relative to that of its competitors. The application of benchmarking concepts generally falls into one of three categories: strategic benchmarking, operational benchmarking, or business management benchmarking. Strategic benchmarking involves the comparison of different business strategies and their potential for success. Through operational benchmarking, a company can understand and attempt to exceed best-in-class companies at a specific activity or function. Operational benchmarking should be used to focus on improving relative cost position or discovering ways to increase market differentiation. Business management benchmarking is the analysis of support functions within competitors of best-of-class firms. Benchmarking is a dynamic process that requires constant rechecking of performance standards. This powerful technique can be used to ensure that a company is steadily improving in customer satisfaction performance.

Bean, T. J. and Gros, J. G. (1992) R&D benchmarking at AT&T. *Research-Technology Management*, **35** (4), 32–37.

As more large technology-based companies use benchmarking to guide their process and quality improvement efforts, they find common concerns. Challenges faced by companies such as AT&T, with many business units working to improve similar processes, result in the need for corporate and local efforts, and mechanisms for sharing benchmarking findings. The authors found that benchmarking is most successful when it is part of a larger quality improvement effort. The value of their benchmarking findings is usually directly correlated with the quality of the long-term relationship with their benchmarking partner, and they often found that the partner is a customer, colleague/collaborator, competitor, or supplier, and often all four requiring much thought on managing the relationship. Their large number of improvement teams select some of the same companies as benchmark partners, requiring coordination and sharing of findings. Much attention is given within AT&T to implementing improvement ideas, often combining top-down and bottom-up approaches.

Bergstrom, A. J. (1992) Business intelligence: a strategic advantage *Bank Marketing*, **24** (10), 28–31.

Having a competitive edge can significantly increase profits and protect against losing business to competitors. Gathering competitor intelligence is one way of gaining an edge. The process of gathering this information includes collecting various kinds of data on current or potential rivals, including financial reports and product-service specifications, then analysing and presenting these data to decision makers. A form of competitor intelligence practiced by some organizations is called benchmarking, which consists of analysing a competitor's product piece by piece or a service delivery procedure by procedure. In most cases, publicly available information is more than adequate for conducting a meaningful analysis. State and local governments can be excellent sources of data. The other side of collecting and analysing information on the competition is protecting, as much as possible, an organization's own product and service delivery secrets.

Betts, M. (1992) Benchmarking helps IS improve competitiveness. *Computerworld*, **26** (48), 1, 20.

Quality benchmarking, not to be confused with hardware performance testing, involves comparing a department's activities with the best practices of other companies. Texas Instruments Inc. has benchmarked such IS topics as data centre costs, help desk operations and managing computer uptime. A comparative study of data centre operations allowed Champion International Corp. to identify cost savings equivalent to 3% of its data centre budget. Many companies reap savings of 20% or more. However, only a minority of IS departments are actively benchmarking. Out of 21 large companies surveyed in Baltimore, Maryland, and Washington, D.C., only four had external benchmarking activities. Recently, 14 companies, including Ford Motor Co., Polaroid Corp., and United Parcel Service Inc., and a federal agency began comparing notes on electronic data interchange. Two consulting firms were brought in – The EDI Group Ltd. and Price Waterhouse – to provide confidential analysis.

Biesada, A. (1992) Strategic benchmarking. *Financial World*, **161** (19), 30–36.

Since 1977, some 86% of the ships that Bath Iron Works delivered were combat vessels for the US Navy. The company never did any contingency planning and was totally unprepared for the end of the Cold War. To search for its future, CEO Duane D. Fitzgerald and his Bath team turned to strategic benchmarking. Instead of asking how a company can excel at a process as in functional benchmarking, strategic benchmarking seeks to incorporate benchmarking into strategic planning. It asks how the company can become world class in tomorrow's probable economic environment. A 1991 survey of 87 member companies of the International Benchmarking Clearinghouse found that while 79% of those polled believed companies will have to benchmark to survive, 95% believed that most companies do not know how to do it. However, as more companies gain experience benchmarking, they may begin to follow Bath Iron Works' path. IBM has used strategic benchmarking at its electronic card assembly and testing facility in Austin, Texas.

Biesada, A. (1991) Benchmarking. *Financial World*, **160** (19), 28–32.

Ed Boyce of Kaiser Associates, which has probably helped more companies benchmark (find and implement best practices) than any other consultancy, says that 60–70% of the largest US companies now have some kind of benchmarking program in place. After getting its start in the manufacturing sector, benchmarking is rapidly making its way into the service and financial service industries and even into government agencies. The approach is to learn from the highest achievers and disseminate that knowledge to all in the company. The organization's total quality management strategy is based on: growth through servicing and originations; a balanced approach to developing complementary businesses; a conservative long-term view toward organizational growth; and dedication to quality. The benefits of best-practices benchmarking include: ensuring the rigor of internal operating targets; creating an openness to new ideas; and raising the organization's perceived level of maximum peak performance. Another early-stage, best-practices approach to encourage people to share success stories was designed by American Residential using an analytic tool developed for the Malcolm Baldrige National Quality Award.

Bracken, D. W. (1992) Benchmarking employee attitudes. *Training & Development*, **46** (6), 49–53.

Many firms have surveyed employee attitudes using paper-and-pencil surveys. Many have regularly compared their survey results to those of other firms. In looking at norms, an organization is comparing itself to an average company. The implication is that it will be satisfied if it can say that it is better than average. This is inconsistent with benchmarking, which involves a company comparing itself to the best companies in a specific performance area. Employee surveys can be a powerful tool as a focal point for communicating and gauging progress toward attaining a corporate vision. The survey process and content must be consistent with the culture that is being created or maintained. To measure progress, a company can design a survey around a vision, mission, values, and goals. By creating its own benchmark, a company can have complete control over the items and response scales and the quality of the data. When managers participate in every phase of the process, they develop a sense of ownership and accountability. Managers should be given the information they need to make informed decisions on examining normative data bases and selecting the options that will work best for the company.

Callam, J. A. (1992) Benchmarking: a tool for continuous improvement. *Human Resource Management*, **31** (1, 2), 141–143.

Benchmarking: A Tool for Continuous Improvement by Kathleen H. J. Liebfried and C. J. McNair is reviewed.

Camp, R. C. (1989) Benchmarking: the search for best practices that lead to superior performance (Part I). *Quality Progress*, **22** (1), 61–68.

Only the approach of establishing targets and productivity programs based on industry best practices leads to superior performance. This process is known as benchmarking. The following basic, philosophical steps of benchmarking are fundamental to success: know the operation; know the industry leaders or competitors; incorporate the best; and gain superiority. Benchmarking is industrial research, or intelligence gathering, that allows managers to compare their functions' performance to the performance of the same function in other companies. Benchmarking identifies those management practices, methods, and processes the function or cost centre would use if it existed in a competitive environment. Benchmarking benefits a company in various ways: it enables the best practices from any industry to be creatively incorporated into the processes of the benchmark function; benchmarking breaks down the reluctance of operations to change; and it may also identify a technological breakthrough that would not have been recognized in one's own industry.

Camp, R. C. (1989) Benchmarking: the search for best practices that lead to superior performance (Part II): Key Process Steps *Quality Progress*, **22** (2), 70–75.

Successful benchmarking centres on achieving several important factors and management behaviours. Management must be committed to making difficult decisions to base operational goals on a concerted view of the external environment. Companies must learn from others and measure themselves regularly against the best in the industry. Creativity in extending the basic process will lead to superior benchmarking results. Benchmarking is a continuous process of measuring against the best; it should be viewed as basically an objective-setting process. Its five steps are: planning what is to be benchmarked and determining to whom or to what the company will be compared; analysing performance gaps and projecting future performance levels; using the findings to set operational targets for change and communicating the findings to all levels; developing action plans and implementing them, including recalibrating the benchmarks if indicated; and attaining the maturity state where the best practices are fully integrated into all business processes.

Camp, R. C. (1989) Benchmarking: the search for industry best practices that lead to superior performance (Part III): Why Benchmark? *Quality Progress*, **22** (3), 76–82.

There are many reasons to benchmark. An understanding of the more important reasons will permit managers to carefully direct their investigations to obtain all of the benefits. The five most important benefits are: end-user requirements are more adequately met; true measures of productivity are determined; goals founded on a concerted view of external conditions are established; a competitive position is attained; and industry best practices are brought into awareness and are sought. Benchmarking is a goal-setting process, but more important, it is a method by which the practices needed to reach the new goals are discovered and understood. Management support and using the process to its fullest extent are both critical to its success. Management involvement is essential because benchmarking directly affects the goals to which management commits. Prior to any benchmarking investigation, there should be a clear understanding of the various work processes in the function.

Camp, R. C. (1989) Benchmarking: the search for industry best practices that lead to superior performance (Part IV). *Quality Progress*, **22** (4), 62–69.

The key to determining what should be benchmarked is to identify the product of the business function. Business processes and practices, physical products, services provided, and the desired level of customer satisfaction can all be benchmarked. In this sense, benchmarking can be used to develop a standard or measure against which to compare. Once the mission statement and logical deliverable outputs have been defined, a solid starting point will have been established to further break down the outputs into specific items to be benchmarked. Current performance measurements may directly indicate what should be benchmarked. The criteria for selecting what to benchmark should be considered early in the formulation of the study. Prior to its benchmarking experience with L. L. Bean, Xerox Corp.'s logistics function had developed a comprehensive mission statement and list of deliverables. What to benchmark was defined as the investigation of the best industry practices for the picking operation and the entire receiving-to-shipping process.

Camp, R. C. (1989) Benchmarking: the search for industry best practices that lead to superior performance (Part V): Beyond Benchmarking *Quality Progress*, **22** (5), 66–68.

Benchmarking has effects that transcend the process itself and its derived benefits, including: management implications; the pursuit of business simplification; and the influence on change initiative. Full use of the benchmarking process, together with other considerations, goes beyond the basics of benchmarking and results in a standard by which to measure the effective simplification of any business. Every function of every business should be considered a candidate for benchmarking. Because of the enormous potential for improvement, there is a critical need for senior management to direct the effort. If benchmarking reveals proven industry best practices that meet but do not exceed customer requirements, then it provides the standard against which to judge the cost-effectiveness of operations. It becomes the vehicle for business simplification, which contributes to improved customer satisfaction and business results. Key considerations that must receive attention to achieve superior performance through benchmarking include: focusing on best practices; meeting end-user requirements; and using quality improvement tools.

References

Aaker, D.A. (1991) *Managing Brand Equity*. The Free Press, New York, USA.

Ammons, J.C. and Gilmour, J.E. (1993) Continuous quality improvement at the Georgia Institute of Technology. *Quality & Productivity Management,* **10**(2), 41–50.

Ballon, R.J. (1992) *Foreign competition in Japan – Human Resource Strategies*. Routledge, London.

Barnet, S. (1992) Strategy and the environment. *The Columbia Journal of World Business,* Fall and Winter, 202–8.

Barthelemy, J.L. (1993) In search of best practice in the management of quality systems audits for continuous improvement: the case of ICL. MBA dissertation, University of Bradford Management Centre, UK.

Benjamin, C. (1993) Honda and the art of competitive manoeuvre. *Long Range Planning*, **26**(4), 22–31.

Booth, J.A. (1993) An assessment of BS5750 within UK organizations. MBA dissertation, University of Bradford Management Centre, UK.

Brown, J. (1992) Competitor analysis – identifying and measuring customer focused performance measures. *Proceedings of benchmarking and competitor analysis in financial services.* 8 December, IIR Ltd, London.

Camp, R.C. (1993) The search for best practices that lead to superior performance. *Pins and Needles*, May–June, 2–7.

Christopher, M. and Yallop, R. (1991) Audit your customer service quality. Cranfield School of Management, Bedford, UK.

Clayton, R. (1992) Strategic partnering: why Alco failed. *Business Quarterly*, Summer, 83–8.

Coate, E. and Maser, J. (1993) O&P manager: TQM initiatives at Oregon State and the University of Pennsylvania. *Quality & Productivity Management*, **10**(2), 11–16.

Cooper, R.G. (1979) Identifying industrial new product success: project NEWPROD. *Industrial Marketing Management*, **8**, 124–35.

Cooper, R.G. (1980) Project NEWPROD: factors in new product success. *European Journal of Marketing*, **14**(5/6), 277–92.

Davies, R. (1992) Improving process management in a customer driven business. *Proceedings of Improved efficiency through benchmarking strategies.* IIR Ltd Industrial Conferences, London.

Day, G.S. (1990) *Market-driven Strategy – Processes for Creating Value.* The Free Press, New York.

Dickson, M. (1992) AT&T plugs a new market. *Financial Times*, 6 Nov.

Dixon, H. (1992) Survey of world industrial review. *Financial Times*, 31 Jan.

Dixon, J.R., Nanni, A.J. and Vollmann, T.E. (1990) *The New Performance Challenge – Measuring Operations for World-class Competition.* Business One Irwin, Illinois.

Drucker, P.F. (1988) Management and the worlds work. *Harvard Business Review*, Sept–Oct, 65–76.

Drucker, P.F. (1954) *The Practice of Management.* Harper and Row, New York.

Feeney, A.R. (1991) Implications of total quality management for health care and the challenges ahead for a NHS pathology department. MBA dissertation thesis, University of Huddersfield, UK.

Forrest, J.E. (1992) Management aspects of strategic partnering. *Journal of General Management*, **17**(4), 25–40.

Fortune, L. (1989) Improving customer–supplier relations in healthcare organizations. *Proceedings of the QQ & PHS quest for quality and productivity in health services*, Washington, USA, 10–13.

Gerlach, M. (1987) Business alliances and the strategy of the Japanese firm. *California Management Review*, Fall, 126–42.

Gray, R. (1989) National Health Service reform. The shape of things to come. *Management Accounting*, 24–5.

Gray, R. and Owen, D. (1992) Environmental reporting award scheme. *Integrated Environmental Management*, **19**.

Greeno, J.L. and Robinson, S.N. (1992) Rethinking corporate environmental management. *The Columbia Journal of World Business*, Fall and Winter, 222–32.

Griffiths, R. (1983) Report of the NHS management inquiry. Department of Health and Social Security, London.

Hellwig, H. (1991) Differences in competitive strategies between the United States and Japan. *IEE Transactions on Engineering Management*, **39**(1), 77–8.

Hideo, I. (1990) *Human Resource Development in Japanese Companies.* Asia Productivity Organization, Tokyo.

Hiromoto, T. (1993) Now it's Japanese accounting. *Pins and Needles*, **19**, May–June, 35–40.

Hudiberg, J.J. (1991) *Winning with Quality – the FPL story.* Quality Resources, USA.

JETRO (1982) *Japanese Corporate Personnel Management.* Japan External Trade Organization, Tokyo, Japan.

Johne, A. and Snelson, P. (1988) Auditing product innovation activities in manufacturing firms. *R&D Management*, **18**(3), 227–33.

Johnson, H.T. (1990) Performance measurement for competitive excellence, in Kaplan, R.S. (ed.) *Measures for Manufacturing Excellence*, Harvard Business School Press, Boston.

Joyer, P.R. (1991) Paul D. Camp Community College. *Quality and Productivity Management*, **9**(2), 61–5.

Kaplan, R.S. (1990) Limitations of cost accounting in advanced manufacturing environments, in Kaplan, R. S. (ed.) *Measures for Manufacturing Excellence*, Harvard Business School Press, Boston.

Kaplan, R.S. and Norton, D.P. (1993) The balanced scoreboard – measures that drive performance. *Quality and Productivity Management*, **10**(3). 47–54.

Koska, M. (1991) New JCAJO standards emphasize continuous quality improvement, *Hospitals*, **65**(15), 41–4.

Kotler, P. and Stonich, P.J. (1991) Turbo marketing through time compression. *The Journal of Business Strategy*, Sept–Oct, 4–29.

Lawton, R.L. (1991) Creating a customer-centred culture in service industries, *Quality Progress*, Sept, 69–72,

Levitt, T. (1983) The globalization of markets. *Harvard Business Review*, May–June, 92–102.

Lewis, J.D. (1992) Competitive alliances redefine companies. *Pins and Needles*, **14**, July–Aug, 2–8.

Loewe, P.M. and Yip, G.S. (1986) *Is it Time to Adopt a Global Strategy?* The MAC Group, Cambridge, Mass.

Lorange, P. and Roos, J. (1991) Why some strategic alliances succeed and others fail. *The Journal of Business Strategy*, Jan–Feb, 25–30.

Maidique, M.A. and Zirger, B.J. (1984) A study of success and failure in product innovation: the case of the US electronics industry. *IEEE Transactions on Engineering Management*, **31**(4), 192–203.

Miller, W. (1991) Gaining competitive advantage through customer satisfaction. *European Management Journal*, **9**(2). 201–12.

Morgan, J. and Morgan, J.J. (1991) *Cracking the Japanese Market – Strategies for Success in the New Global Economy*. The Free Press, New York.

Motteram, G.J. (1991) Paying the ferryman. *Manufacturing Engineer*, Dec 1990 – Jan 1991, 33–5.

Naumann, E. and Shannon, P. (1992) What is customer-driven marketing? *Business Horizons*, July–Aug, 45–51.

Newman, J.C. and Breeden, K.M. (1992) Lessons from environmental leaders. *Columbia Journal of World Business*, Fall and Winter, 210–21.

Ohmae, K. (1982) *The Mind of the Strategist – the Art of Japanese Business*. McGraw-Hill Book Company, New York, USA.

Ohmae, K. (1991) *Getting Back to Strategy; Seeking and Seeing Competitive Advantage*. Harvard Business Review Book Series.

Ozawa, T. (1982) *People and Productivity in Japan*. Pergamon Press, New York, USA.

Pascale, R.T. and Athos, A.G. (1982) *The Art of Japanese Management*. Penguin, Harmondsworth.

Peters. T.J. and Waterman, R.H. (1982) *In Search of Excellence*. Harper and Row, New York.

Polonsky, M. and Zeffare, R. (1992) Corporate environmental commitment in Australia: a sectoral comparison. *Business Strategy and the Enviroment*, **1**(2).

Priestly, T. (1993) Getting benchmarking right from the beginning and keeping it on track. *Proceedings of Achieving best practice through practical benchmarking*. IIR Ltd Industrial Conferences, London.

Redman. N. (1993) Benchmarking to support a vision. *Proceedings of Benchmarking for strategic advantage conference*, London 6 July, ICM Marketing Ltd.

Rhinesmith, S.H. (1991) Going global from the inside out. *Training and Development*, 43–7.

Robert, M. (1992) The do's and don'ts of strategic alliances. *The Journal of Business Strategy*, March–April, 50–3.

Rothwell, R. *et al.* (1974) *SAPPHO updated*. Science Policy Research Unit, Brighton.

Ryan, M. (1988) Assessment: the first step in image management. *Tokyo Business Today*, September, 36–8.

Saunders, M. (1993) Improving customer service – benchmarking what matters to the customer. *Proceeding for Strategic advantage conference*, London, 6 July, ICM Marketing Ltd.

Saxton, J. and Locander, W.B. (1991) A systems view of strategic planning at Procter and Gamble, in *Competing Globally through Customer Values* (eds M.J. Stahl, and G.M. Bounds), Quorum Books, New York, USA.

Shapiro, B.P. (1988) What the hell is market-orientated? *Harvard Business Review*, Nov–Dec, 119–25.

Simon, F.L. (1992) Marketing green products in the triad. *The Columbia Journal of World Business*, Fall and Winter, 269–85.

Sink, D.S. (1993) Quality priorities of higher education. *Quality and Productivity Management*, **10**(2), 5–6.

Souder, W.E. (1987) *Managing New Product Innovations*. Lexington Books, USA.

Spanner, G.E., Nuno, J.S. and Chandra, C. (1993) Time-based strategies – theory and practice. *Long Range Planning*, **26**(4), 90–101.

Stalk, G. and Hout, T.M. (1990) *Competing against Time – how Time-based Competition is reshaping Global Markets*. The Free Press, New York.

Starr M.K. (1992) Accelerating innovation. *Business Horizons*, July–Aug, 45–51.

Sterne, D. (1992) Core competences: the key to corporate advantage. *Multinational Business*, **3**, 102–19.

Stikker, A. (1992) Sustainability and business management. *Business Strategy and the Environment*, **1**(3), 1–8.

Taylor, P. (1992) Survey of international telecommunications. *Financial Times*, 15 Oct.

Theuerkauf, I. (1991) Reshaping the global organization. *The McKinsey Quarterly*, **3**, 102–19.

Turpin, D. (1993) Strategic alliances with Japanese firms: myths and realities. *Long Range Planning*, **26**(4), 11–15.

Umachonu, V.K. (1991) *Total Quality Productivity Management in Healthcare Organizations*. Industrial Engineering and Management Press, Institute of Industrial Engineers, USA.

Waterman, R.H., Peters, T.J. and Phillips. J.R. (1980) Structure is not organization. *Business Horizons*, June 14–26.

Watson, G.H. (1993) *Strategic Benchmarking: How to Rate your Company's Performance against the World's Best*, John Wiley & Sons Inc, New York.

Westcott II. W.F. (1992) Environmental technology cooperation – a quid pro quo for transnational corporations and developing countries. *The Columbia Journal of World Business*, Fall and Winter, 144–53.

Wurster T.S. (1987) *The Leading Brands: 1925–1985*. Perspectives, The Boston Consulting Group.

Williamson, J. (1992) Survey of international telecommunications. *Financial Times*, 15 Oct.

Zairi, M. (1990) The management of advanced manufacturing technology: a study of user–supplier networks. PhD dissertation, The management centre, University of Staffordshire, England.

Zairi, M. (1991) *Total Quality Management for Engineers*. Woodhead Publishing Ltd., UK.

Zairi, M. (1992) *Management of Advanced Manufacturing Technology*. Sigma Press, Wilmslow, UK.

Zairi, M. (1992) *Competitive Benchmarking – an Executive Guide*, Technical Communications (Publishing) Ltd. UK.

Index

GPSR Compliance

The European Union's (EU) General Product Safety Regulation (GPSR) is a set of rules that requires consumer products to be safe and our obligations to ensure this.

If you have any concerns about our products, you can contact us on ProductSafety@springernature.com

In case Publisher is established outside the EU, the EU authorized representative is:

Springer Nature Customer Service Center GmbH
Europaplatz 3
69115 Heidelberg, Germany

Batch number: 10761599

Printed by Printforce, the Netherlands